The Archaean: Geological and Geochemical Windows into the Early Earth

Modern Approaches in Solid Earth Sciences

VOLUME 9

Series Editors

Yildirim Dilek, *Department of Geology and Environmental Earth Science, Miami University, Oxford, OH, U.S.A.*
Franco Pirajno, *Geological Survey of Western Australia, and The University of Western Australia, Perth, Australia*
M.J.R. Wortel, *Faculty of Geosciences, Utrecht University, The Netherlands*

For further volumes:
http://www.springer.com/series/7377

Andrew Y. Glikson

The Archaean: Geological and Geochemical Windows into the Early Earth

Andrew Y. Glikson
School of Archaeology and Anthropology
  and Planetary Science Institute
Australian National University
Canberra, Australian Capital Territory
Australia

Responsible Series Editor: F. Pirajno

Additional material to this book can be downloaded from http://extras.springer.com

ISSN 1876-1682　　　　　　　　ISSN 1876-1690 (electronic)
ISBN 978-3-319-07907-3　　　　ISBN 978-3-319-07908-0 (eBook)
DOI 10.1007/978-3-319-07908-0
Springer Cham Heidelberg New York Dordrecht London

Library of Congress Control Number: 2014945202

© Springer International Publishing Switzerland 2014
This work is subject to copyright. All rights are reserved by the Publisher, whether the whole or part of the material is concerned, specifically the rights of translation, reprinting, reuse of illustrations, recitation, broadcasting, reproduction on microfilms or in any other physical way, and transmission or information storage and retrieval, electronic adaptation, computer software, or by similar or dissimilar methodology now known or hereafter developed. Exempted from this legal reservation are brief excerpts in connection with reviews or scholarly analysis or material supplied specifically for the purpose of being entered and executed on a computer system, for exclusive use by the purchaser of the work. Duplication of this publication or parts thereof is permitted only under the provisions of the Copyright Law of the Publisher's location, in its current version, and permission for use must always be obtained from Springer. Permissions for use may be obtained through RightsLink at the Copyright Clearance Center. Violations are liable to prosecution under the respective Copyright Law.
The use of general descriptive names, registered names, trademarks, service marks, etc. in this publication does not imply, even in the absence of a specific statement, that such names are exempt from the relevant protective laws and regulations and therefore free for general use.
While the advice and information in this book are believed to be true and accurate at the date of publication, neither the authors nor the editors nor the publisher can accept any legal responsibility for any errors or omissions that may be made. The publisher makes no warranty, express or implied, with respect to the material contained herein.

Printed on acid-free paper

Springer is part of Springer Science+Business Media (www.springer.com)

The book is dedicated to the memory of Alec Trendall, pioneer of Precambrian research, as well as a tribute to the contributions to Archaean field and isotopic geology and asteroid impact research by Arthur Hickman, Don Lowe, Allen Nutman, Bruce Simonson, Martin Van Kranendonk, Gary Byerly, Bob Pidgeon, Bill Compston, John Valley, Simon Wilde, Peter Haines and the late Ian Williams.

Vestiges of a Beginning

> In dealing with rocks formed when the world was less than half of its present age, a strict adherence to the doctrine of uniformitarianism is considered unjustified. (Macgregor 1951, p. 27)

Determinations of the age of the Earth at 4.54±0.05 billion years, based on radiometric U-Pb radiometric age of meteorites and consistent with the 4.567 Ga age of the solar system, as determined from Ca-Al inclusions in carbonaceous chondrites, lunar samples and the oldest measured terrestrial zircons <4.404 Ga-old from the Jack Hills, Western Australia, leave large part of the earliest history of the planet unknown. Where the zircons signify relic felsic rocks as far back as ~4.4 Ga, due to selective retention of zircons and quartzite and destruction of more labile lithologies, i.e. mafic and ultramafic rocks, the composition of the bulk of the Earth crust at that stage remains unknown. Isotopic and geochemical signatures in rocks as old as ~3.8 Ga indicate an evolutionary trend from mafic-ultramafic crust to tonalite-trondhjemite-granodiorite (TTG)-dominated micro continental nuclei. To date, signatures of the 3.95–3.85 Ga Late heavy Bombardment (LHB), manifested by the lunar Mare, have not been discovered on Earth. ~4.0–3.8 Ga high-grade metamorphic terrains, known from Slave province and Antarctica; ~3.8 Ga supracrustal volcanic-sedimentary belts represented by the Isua and Akilia terrains in southwest Greenland; and mafic rocks from the northern Superior Province may conceal evidence for the LHB. Recent discoveries of near to 14 Archaean impact ejecta units up to 3.48 Ga-old intercalated with volcanic and sedimentary rocks in the Barberton and Pilbara greenstone belts, including clusters about 3.25–3.22 Ga and 2.63–2.48 Ga in age, may represent terrestrial vestiges of an extended LHB. The interval of ~3.25–3.22 Ga ago emerges as a major break in Archaean crustal evolution when, as indicated by a major body of field and isotopic age evidence, major asteroid bombardment resulted in faulting, large scale uplift, intrusion of granites and an abrupt shift from crustal conditions dominated by mafic-ultramafic crust associated with emplacement of tonalite-trondhjemite-granodiorite (TTG) plutons, to semi-continental nuclei represented by arenites, turbidites, conglomerate, banded iron formations and felsic volcanics. At this stage, pre-3.2 Ga dome-structured granite-greenstone systems were largely replaced by linear accretional granite-greenstone systems such as the Superior Province in Canada, Yilgarn Craton and the western Pilbara Craton, compared by some authors to circum-Pacific arc-trench settings. A fundamental geotectonic transformation is consistent with the increasing role of garnet fractionation as indicated by Al-depleted and plagioclase-enriched magmatic compositions, suggesting cooler high P/T (pressure/temperature) mantle and crustal magma sources, consistent with development of subduction. A concentration of large impacts during 2.63–2.48 Ga potentially accounts for peak magmatic events culminating the Archaean era. Strict comparisons between the Archaean systems and modern Arc-trench geotectonic setting will be shown to be unwarranted. These observations are considered in this monograph in the context of a review of major aspects of the Archaean record, in particular in the southern continents but with reference to northern cratons. The book provides an excursion through granite-greenstone terrains, and to a lesser extent high-grade metamorphic terrains, focusing on relic primary features including volcanic, sedimentary, petrological, geochemical and paleontological elements, with the aim of elucidating the nature of original environments and processes which dominated environments in which early life forms have emerged. By contrast to uniformitarian models, which take little or no account of repeated impacts of large asteroid clusters and their effects during ~3.47–2.48 Ga, the Archaean geological record will be shown to be consistent with the theory of asteroid impact-triggered geodynamic and magmatic activity originally advanced by D. H. Green in 1972 and 1981.

Acknowledgements

This book conveys my observations in Archaean terrains following my studies at the University of Western Australia during 1964–1968, my years in the Australian Geological Survey (BMR/AGSO/GA) during 1968–1996 and the Australian National University to the present. The list of people I like to thank is long; however I particularly like to thank those who have given me special collaboration, advice and support: John Vickers, for many years of collaboration in geological field work in Western Australia; The late Alec Trendall and Tom Vallance, for their generous petrological advice during 1965–1967; Bill O'Beirne, Jack Hallberg, Gary Robinson and Kingsley Mills, for help with my Ph.D. thesis; Arthur Hickman, Franco Pirajno, the late Ian Williams, Peter Haines and Martin Van Kranendonk in connection with collaboration with the Geological Survey of Western Australia; the late Shen Su Sun, John Sheraton, Ian Lambert, Doone Wyborn, John Ferguson, Alan Whittaker, John Casey, the late Walter Dallwitz and Chris Pigram during my years in the BMR/AGSO (now Geoscience Australia); David Green, Ian Jackson, Richard Arculus, Chris Klootwijk and Prame Chopra for support at the Research School of Earth Science, Australian National University. I thank my overseas colleagues for collaboration, in particular Lutz Bischoff, Alec Baer, Don Lowe, Gary Byerly, Bruce Simonson, Scott Hassler, and Shigenori Maruyama. I am grateful to John Ferguson, Miryam Glikson, Victor Gostin, Arthur Hickman, Gerta Keller, Victor Masaitis, Franco Pirajno and Bruce Radke for reviews and comments on this book manuscript and to Brenda McAvoy for meticulous proof reading. I thank G. Byerly, Arthur Hickman, Michele McLeod, Reg Morrison, J-F Moyen, Allen Nutman, Bill Schopf, Bruce Simonson and John Valley, Geoscience Australia and the Geological Survey of Western Australia for permission to reproduce figures and photographic material. I thank Franco Pirajno, Petra van Steenbergen and Corina Van der Giessen for their advise and editorial assistance.

Contents

1	**The Moon and the Late Heavy Bombardment (LHB)**................	1
2	**Hadean and Early Archaean High Grade Metamorphic Terrains**................	9
	2.1 Mount Narryer >4.0 Ga Detrital Zircons........................	11
	2.2 Acasta Gneiss and Hadean Zircons.............................	14
	2.3 Northeastern Superior Province Gneisses.................	16
	2.4 Wyoming Gneisses...	17
	2.5 Itsaq Gneisses and Supracrustals, Southwest Greenland...........	18
	2.6 Siberian Gneisses...	22
	2.7 North China Craton...	22
	2.8 Antarctic Gneisses..	23
3	**Early Archaean Mafic-Ultramafic Crustal Relics**...............	25
4	**The Tonalite-Trondhjemite-Granodiorite (TTG) Suite and Archaean Island Continents**................	35
	4.1 Archaean Batholiths..	36
	4.2 Vertical Crustal Zonation.....................................	40
5	**Isotopic Temporal Trends of Early Crustal Evolution**............	43
	5.1 U-Th-Pb Isotopes..	43
	5.2 Rb-Sr Isotopes..	45
	5.3 The Lu-Hf System..	46
	5.4 The Sm-Nd System...	48
	5.5 The Re-Os System..	51
6	**Geochemical Trends of Archaean Magmatism**.....................	53
	6.1 Mafic and Ultramafic Volcanics...............................	54
	6.2 Felsic Igneous Rocks...	60

7	**Pre-3.2 Ga Evolution and Asteroid Impacts of the Barberton Greenstone Belt, Kaapvaal Craton, South Africa**	73
	7.1 The Barberton Greenstone Belt, Eastern Kaapvaal Craton	73
	7.1.1 Onverwacht Group	81
	7.2 Evolution of the Zimbabwe Craton	86
	7.3 Pre-3.2 Ga Asteroid Impact Units of the Kaapvaal Craton	88
8	**Evolution and Pre-3.2 Ga Asteroid Impact Clusters: Pilbara Craton, Western Australia**	97
	8.1 Crustal Evolution	97
	8.2 ~Pre-3.2 Ga Impact Fallout Units in the Pilbara Craton	110
	8.3 ~3.24–3.227 Ga Impact-Correlated Units	114
9	**Post-3.2 Ga Granite-Greenstone Systems**	119
	9.1 Evolution of the Yilgarn Craton	119
	9.2 Evolution of South Indian Granite-Greenstone Terrains	126
	9.3 Accretion of Superior Province Terrains	129
	9.4 Archaean Fennoscandian/Baltic Terrains	130
10	**Post-3.2 Ga Basins and Asteroid Impact Units**	131
	10.1 Late Archaean Griqualand West Basin	131
	10.2 Late Archaean Fortescue and Hamersley Basins Impact Units	135
	10.2.1 ~2.63 Ga Jeerinah Impact Layer and the Roy Hill and Carawine Dolomite Impact/Tsunami Megabreccia	135
	10.2.2 ~2.57 Ga Paraburdoo Impact Spherule Unit (PSL)	144
	10.2.3 ~2.56 Ga Spherule Marker Bed (SMB)	144
	10.2.4 ~2.48 Ga Impact, Dales Gorge (DGS4)	151
	10.2.5 Estimates of Asteroid Size and Compositions	154
	10.3 Impact Fallout Units and Banded Iron Formations	156
	10.4 Inter-continental Correlation of Impact Units	157
11	**The Early Atmosphere and Archaean Life**	159
	11.1 Archaean Carbon-Oxygen-Sulphur Cycles and the Early Atmosphere	160
	11.2 Archaean Life	170
12	**Uniformitarian Theories and Catastrophic Events Through Time**	177
Appendices		185
About the Author		193
References		195
Index		225

Chapter 1
The Moon and the Late Heavy Bombardment (LHB)

Abstract The Lunar surface offers the best evidence for events since accretion of the Earth-Moon system during the pre-Nectarian (pre-3.9 Ga) to about 3.85 Ga (Nectarian) whereas the terrestrial records for this era, referred to as the Hadean, are limited to high-grade metamorphic terrains (to ~4.0 Ga) and detrital or xenocrystic zircons (to ~4.4 Ga). Asteroid bombardment of the Moon during ~3.95–3.85 Ga (Late Heavy Bombardment – LHB), forming the lunar Mare, has not to date been identified in the terrestrial records. Interpretations of the terrestrial zircon record in terms of a Hadean felsic continental crust suffer from a major sampling bias in favor of preservation of resistant zircon grains vs an absence of labile mafic detrital material, rendering the overall composition of the pre-4.0 Ga Earth crust uncertain. Lunar impacts associated with Mare volcanism about ~3.2 Ga correlate with a large impact cluster which affected Archaean mafic-ultramafic crust and associated TTG (tonalite-trondhjemite-granodiorite) plutons, triggering abrupt uplift of the greenstone-TTG crust, forming small continental nuclei documented in the eastern Kaapvaal Craton, South Africa, and the Pilbara Craton, northwestern Australia. As suggested by Lowe and Byerly (Did the LHB end not with a bang but with a whimper? 41st Lunar Planet Science conference 2563pdf, 2010), the Archaean asteroid bombardment can be regarded as an extension of the LHB on Earth.

Keywords Moon • Late Heavy Bombardment • Lunar mare • Lunar spherules • Mare volcanism • Ti-rich basalt

Early terrestrial beginnings are interpreted in terms of a cosmic collision between an embryonic semi-molten Earth and a Mars-scale body—Theia ~4.5 billion years-ago determined from Pb isotopes (Stevenson 1987). The consequent formation of a metallic core, inducing a magnetic field which protects the Earth from cosmic radiation, and a strong gravity field which retards atmospheric gases from escaping into space, resulted in a haven for life at the Earth's surface (Gould 1990). It is assumed that, by analogy to the Lunar history (Figs. 1.1, 1.2, and 1.3), Earth was subject to

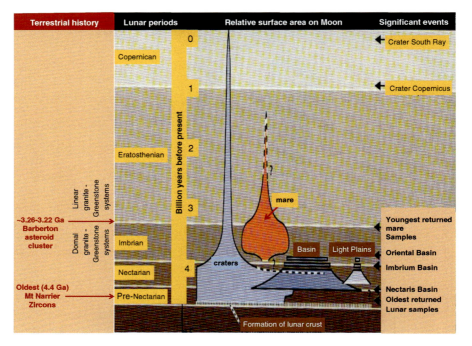

Fig. 1.1 Broad stratigraphy of the Moon with the early Earth, marking the distribution of the lunar basins, light plains, craters and mare, correlated with Archaean stages and events on Earth (Data source: NASA)

heavy bombardment by asteroids during ~3.95–3.86 Ga (Ryder 1990, 1991, 1997). Discoveries of detrital and xenocrystic zoned zircons from Mount Narryer and Jack Hills (Western Australia) (Fig. 2.1), containing ~4.4 Ga cores and younger zones (Fig. 2.2) represent the only direct record of the terrestrial Hadean era (pre-4.0 Ga), a term coined by Cloud (1972). The zircons signify vestiges of felsic magmatic rocks, with high $\delta^{18}O$ values (>5 ‰ VSMOW) (Valley et al. 2002) implying relatively low temperatures and thereby a presence of a surface water component during alteration of the zircons (Wilde et al. 2001; Mojzsis et al. 2001; Valley et al. 2002, 2006; Valley 2005) (Figs. 11.2 and 11.3).

Early Precambrian terrains contain relict <4.1 Ga-old supracrustal and plutonic components, exposed in Greenland, Labrador, Slave Province, Minnesota, Siberia, northeast China, southern Africa, India, Western Australia and Antarctica (Van Kranendonk et al. 2007a, b), although in some instances U-Pb zircon ages may be inherited. Amphibolite to granulite grade metamorphism of these formations, some of which formed parallel to the Late Heavy Bombardment (LHB) on the Moon (~3.95–3.85 Ga) (Ryder 1990), obscures recognition of primary features, precluding identification of signatures of the LHB. During the LHB exposure of the Earth surface to cosmic and UV radiation, incineration affected by large asteroid impacts and acid rain, likely restricted or precluded photosynthesis (Zahnle and Sleep 1997; Chyba 1993; Chyba and Sagan 1996). On the other hand extremophile chemotrophic

1 The Moon and the Late Heavy Bombardment (LHB)

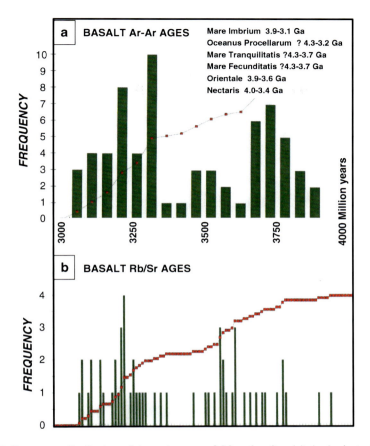

Fig. 1.2 Frequency distribution of isotopic ages of Mare basalts: (**a**) Ar-Ar isotopic ages; (**b**) Rb-Sr isotopic ages, based on data from Basaltic Volcanism of the Terrestrial Planets (BVSP 1981)

bacteria (the *'deep hot biosphere'* of Gold 1999) are likely to have resided in faults and fractures following formation of original crustal vestiges. The suggestion that low $\delta^{13}C$ graphite clouding in apatite in ~3.85 Ga-old banded iron formation (BIF) in south-western Greenland may represent biogenic effects (Mojzsis and Harrison 2000) has not been confirmed, as the clouding was shown to arise from secondary contamination (Nutman and Friend 2006). By contrast, new evidence has emerged for a biogenic-mediated deposition of dolomite and possibly of BIF in the Isua greenstone belt (Bolhar et al. 2004; Nutman et al. 2010).

The LHB has been alternatively interpreted in terms of the tail-end of planetary accretion or as a temporally distinct bombardment episode (Ryder 1990, 1991). Mare volcanism is recorded for a period of about 1.3 billion years (~3.8–~2.5 Ga). Ar-Ar and Rb-Sr isotopic age frequencies indicate peak volcanic activities about ~3.8–3.7 Ga and ~3.4–3.2 Ga (Fig. 1.2). A ~3.8–3.5 Ga (Imbrian) Ti-rich mare

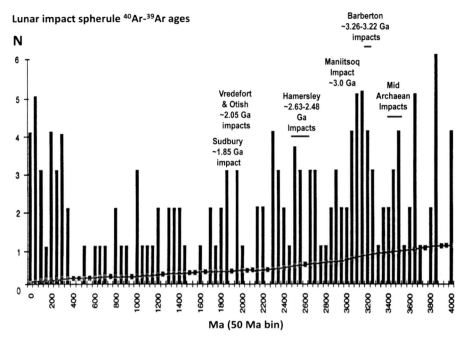

Fig. 1.3 Frequency distribution frequency of $^{40}Ar/^{39}Ar$ ages of Lunar impact glass spherules from the lunar soil, based on data by Culler et al. (2000)

volcanic phase (Apollo 11 and Apollo 17 type basalts) flooded large areas of the eastern portion of the lunar near-side (Head 1976). Eruption of ~3.5–3.0 Ga (Middle to late Imbrian) less Ti-rich basalts (Apollo 12 and Apollo 15) flooded widespread areas of the moon. About ~3.0–2.5 Ga (early Eratosthenian) Ti-rich basalts flooded portions of Mare Imbrium and the western Mare (Head 1976). Some of the largest lunar Mare basins contain low-Titanium basalts which likely represent impact-triggered volcanic activity (Ryder 1997), including Mare Imbrium (3.86 ± 0.02 Ga) and associated 3.85 ± 0.03 Ga-old K, REE, and P-rich-basalts (KREEP) (BVSP 1981). Similar relationships between impact and volcanic activity may pertain in Oceanus Procellarum (3.29–3.08 Ga basalts) and in Hadley Apennines (3.37–3.21 Ga basalts) (BVSP 1981). The likelihood of impact-volcanic relationships of the lunar mare gains support from laser $^{40}Ar/^{39}Ar$ analyses of lunar impact spherules from sample 11199 (Apollo 14, Fra Mauro Formation) (Culler et al. 2000; Levine et al. 2005; Hui et al. 2009) (Fig. 1.3).

Impact events tentatively indicated by the lunar spherule data include 3.87 Ga, 3.83 Ga, 3.66 Ga, 3.53 Ga and 3.47 Ga peaks. A significant age spike is indicated at 3.18 Ga, i.e. near the boundary between the Late Imbrian lunar era (3.9–3.2 Ga) and the post-Mare Eratosthenian lunar era (3.2–1.2 Ga), as defined by the cratering record (Wilhelms 1987). Some 34 lunar impact spherules yield a large-error mean age of 3,188 ± 198 Ma with a median age at 3,181 Ma, whereas 7 spherule ages with error <100 m.y. yield a large-error mean age of 3,178 ± 80 Ma and a median at 3,186 Ma (Fig. 1.3), which correlates approximately with the 3.256–3.225 Ga impact cluster indicated by impact condensate spherule units in the Barberton Greenstone Belt, Transvaal, South Africa (Figs. 7.9–7.17).

Despite the occurrence of >3.8 Ga rocks (Chap. 2) to date no direct evidence of the effects of the LHB on early Archaean crust has been detected (Koeberl 2006). The penetratively deformed high-grade metamorphic state of >3.8 Ga terrains overprints and obscures primary textural and mineralogical features diagnostic of shock metamorphism. Nor have shock metamorphic (planar deformation features) been detected in zircons and quartz grains entrained as xenocrysts in magmatic and metamorphic rocks or in detrital sediments, a likely explanation being the structurally weaker state and thus destruction of shock metamorphosed minerals. Schoenberg et al. (2002) reported tungsten isotope anomalies based on the ^{182}Hf-^{182}W system in metamorphosed sedimentary rocks from the 3.8–3.7-Ga Isua greenstone belt, West Greenland and northern Labrador. The short half-life of 9 Ma of the parent ^{182}Hf isotope and the preservation of the ^{182}W daughter isotope imply pre-existence of components of the primordial solar system in the metamorphosed sediments. Jorgensen et al. (1999) report an average Iridium level of 150 ppt from clastic and banded iron sediments in southwest Greenland, about seven times higher than present day ocean crust (20 ppt), an enrichment attributed by these authors to cometary impacts. Examination of the ^{53}Mn-^{53}Cr and ^{53}Cr-^{52}Cr isotopic compositions of metamorphosed turbidite, pelagic sediments and banded ironstones in the ~3.7 Ga Isua Supracrustal Belt, southwest Greenland, did not detect meteoritic components (Frey and Rosing 2005). According to these authors unequivocal evidence of an LHB remains uncertain. Possible explanations include unrepresentative samples and sedimentation period which did not overlap with asteroid impacts. Potential chromium anomalies, if present, may be too small to be detected.

Tentative correlations between the lunar spherule data and terrestrial events include the following (Fig. 1.3):

(A) Sharp ~3.53 Ga and ~3.47 Ga lunar maxima closely correlate with peak magmatic activity associated with volcanic activity in Archaean greenstone–granitoid systems in the Pilbara Craton, Western Australia, and Kaapvaal Craton, Transvaal;

(B) A ~1.8 Ga lunar peak, including spherule ages of 1,813 ± 36.3 Ma, occurs within error from the 1.85 Ga age of the >250 km-diameter Sudbury impact structure, Ontario (Grieve 2006);

(C) Two of the shallower lunar age peaks overlap the age of impact spherules in the Hamersley Basin, Western Australia, including ~2.56–2.57 Ga spherule units in the Wittenoom Formation (Sect. 10.2.3) and a ~2.48 Ga spherule unit in banded ironstones of the Dale Gorge Member (Simonson 1992; Simonson and Hassler 1997; Simonson and Glass 2004);

(D) The ~1.08 Ga peak parallels the emplacement of the large layered mafic-ultramafic Giles Complex, central Australia (Ballhaus and Glikson 1995; Glikson 1995).

Nine lunar spherule ages, with errors of less than 20 Ma, allow comparisons with terrestrial isotopic ages of similar or better precision (Fig. 1.3). These include (1) 3,872.1 ± 11.2 Ma and 3,807.4 ± 14.5 Ma lunar spherule ages corresponding in part to the LHB; (2) 3,542 ± 11.4 and 3475.8 ± 15 Ma ages correlated in part with Archaean volcanic events and impacts in the Pilbara and Kaapvaal cratons, and (3) 3,186.2 ± 12.1, 352.6 ± 6.6 (the latter corresponding to the end-Devonian), 304.6 ± 17 and 51.4 ± 4.0 Ma ages. Notable in their absence are spherule age peaks corresponding to the ~2.023 Ga ~300 km-diameter Vredefort impact structure and the ~500 km Otish Basin impact (Genest et al. 2010). The lunar spherule age data reflect the late Devonian impact cluster (Woodleigh, Western Australia; Mory et al. 2000), D = 120 km, 364.8 Ma (Uysal et al. 2001), Charlevoix, Quebec, 367.15, D = 54 km; Siljan, Sweden, 368.11 Ma, D = 52 km; Alamo breccia, Nevada, Frasnian-Fammenian age, including impact spherules, shocked quartz and Ir anomalies; Ternovka, Ukraine, 350 Ma, D = 15 km; Kaluga, Russia, 380.10, D = 15 km; Ilynets, Ukraine, 395.5 Ma, D = 4.5 km; Elbow, Saskatchewan, 395.25 Ma, D = 8 km) and contemporaneous Frasnian-Fammenian and end-Devonian mass extinctions. The late Devonian events may be mirrored by a precise lunar spherule age of 352.6 ± 6.6 Ma age and a cumulative peak of 11 spherule ages with errors >100 m.y.. Other potential correlations may be indicated by lunar spherule ages include ~303 Ma (Carboniferous–Permian boundary – Warburton twin impacts – Glikson 2013), ~251 Ma (Permian–Triassic boundary), and ~103 Ma, 64–71 Ma (Cretaceous–Tertiary boundary), ~52 and ~38–32 Ma (late Eocene impacts and extinction).

Most Archaean crustal models have neglected the consequences of impact by large asteroids in the inner solar system in the wake of the LHB, assuming a decline in flux from 4 to 9.10^{-13} km^{-2} year^{-1} (for craters Dc >=18 km) to a flux of 3.8–$6.3.10^{-15}$ km^{-2}.year^{-1} (for craters Dc >=20 km) (Grieve and Dence 1979; Baldwin 1985; Ryder 1991). Post-LHB impact rates estimated from lunar and terrestrial crater counts are of the same order of magnitude as the cratering rate, i.e. $5.9 \pm 3.5.10^{-15}$ km^{-2} year^{-1} (for craters Dc >=20 km) estimated from astronomical observations of near-Earth asteroids (NEA) and comets (Shoemaker and Shoemaker 1996). Combining these rates with crater size vs cumulative size frequency relationships approximating $N \infty Dc^{-1.8}$ to $N \infty Dc^{-2.0}$ (Dc = crater diameter; N = cumulative number of craters with diameters larger than Dc), implies the formation of more than 53 craters of Dc >= 300 km, and more than 19 craters of Dc >= 500 km since ~3.8 Ga, nearly two orders of magnitude higher than the preserved record. The size distribution relationship $N \infty D^{-1.8}$ is probably only valid for craters <100 km. NEAs in the range of Dp = 0.1–5.0 km correspond better to the relation $N \infty Dp^{-2.0}$ consistent with young cratered surfaces on the terrestrial planets (Glikson 2001). The frequency of impacts by asteroids the size of Eros (long axis Dp = 34 km), the comet Swift Tuttle (Dp = 26 km) and Ganymede (Dp = 32–34 km) is uncertain, but a preliminary estimate suggests at least 12 impacts by bodies of this scale since the LHB.

Comparing estimates of post-LHB impacts with the preserved terrestrial cratering record documented to date, the minimum extraterrestrial impact incidence for craters Dc >=100 km is defined by the eight observed impact structures of Dc >=100 km, including Maniitsoq – ~300 km; Vredefort – 298 km; Sudbury – 250 km;

Chicxulub – 170 km; Woodleigh – 120 km; Manicouagan – 100 km; Popigai – 100 km; Warburton >400 km (Glikson et al. 2013) and Otish Basin of possibly ~500 km diameter (Genest et al. 2010), which struck continental crust. Based on the Lunar impact flux a minimum number of 30 continental craters of Dc > 100 km and ten continental craters of Dc >= 250 km may be estimated.

The Lunar LHB evidence records a rapid decline of the magnitude and rate of impacts following ~3.85–3.80 Ga (Cohen et al. 2000; Hartmann et al. 2000). By contrast, since the discovery of ~3.48–3.22 Ga asteroid impact fallout units in the Barberton Greenstone Belt, South Africa (Lowe et al. 1989; Lowe et al. 2003) and in ~2.63–2.48 Ga sediments of the Hamersley Basin, Pilbara Craton, Western Australia (Simonson 1992), more than 14 impact fallout/ejecta units were identified in Archaean terrains, including:

- Barberton Greenstone Belt (Lowe et al. 2003; Lowe and Byerly 2010): 3,482 Ma, 3,472 Ma, 3,445 Ma, 3,416 Ma, 3,334 Ma, 3,256 Ma, 3,243 Ma, 3,225 Ma (2 units).
- Transvaal Basin: 2,647 ± 30 Ma (Monteville Spherule Layer), 2,581 ± 9 Ma (Reivilo Spherule Layer), 2,516 ± 4 Ma (Kuruman Spherule layer).
- Pilbara Craton and Hamersley Basin, Western Australia (Simonson and Glass 2004; Glikson 2013): ~3.47 Ga (2 fallout units), ~2.63 Ga, ~2.57 Ga, ~2.56 Ga (2 fallout units), ~ 2.48 Ga.

Another Archaean impact structure is documented in southwestern Greenland by the ~2.975 Ga Maniitsoq impact structure (Garde et al. 2012). Based on the above evidence, Lowe and Byerly (2010) suggest the LHB continued through the early Archaean, an idea supported by a view of the Archaean impact record as the '*tip of the iceberg*' in terms of the limited preservation of impact fallout units and the difficulty in their identification in the field. Due to the incomplete nature of the Archaean sedimentary record, large parts of which have been removed as evidenced by unconformities and paraconformities, likely less than 10 % of asteroid impact ejecta units have been detected to date in Archaean and Proterozoic terrains. This potentially opens the door to a paradigm shift in terms of the understanding of early crustal evolution.

Chapter 2
Hadean and Early Archaean High Grade Metamorphic Terrains

> *Who knows for certain*
> *Who shall here declare it*
> *Whence was it born, whence came creation*
> *The Gods are later than this world's formation*
> *Who then can know the origins of the world*
>
> (The Rig Veda, X.129).

Ancient Water / Marble Bar
No one
Was there to hear
The muffled roar of an earthquake,
Nor *anyone* who froze with fear
Of rising cliffs, eclipsed deep lakes
Sparkling comet-lit horizons
Brighter than one thousand suns
That blinded *no one's* vision.
No one
Stood there in awe
Of an angry black coned volcano
Nor any pair of eyes that saw
Red streams eject from inferno
Plumes spewing out of Earth
And yellow sulphur clouds
Choking *no one's* breath.
No one
Was numbed by thunder
As jet black storms gathered
Nor anyone was struck asunder
By lightning, when rocks shuttered
Engulfed by gushing torrents
That drowned smouldering ashes
Which *no one* was to lament.

In time
Once again an orange star rose
Above a sleeping archipelago
Sun rays breaking into blue depth ooze
Waves rippling sand's ebb and flow
Receding to submerged twilight worlds
Where budding algal mats
Declare life
On the young Earth.

A poem by Andrew Glikson

Abstract The oldest terrestrial minerals identified to date are 4,363±20 Ma-old detrital zircons from Jack Hills, Western Australia (Valley et al. Geology 36:911–912, 2002; Nemchin et al. Earth Planet Sci Lett 244:218–233, 2006; Pidgeon and Nemchin, Precamb Res 150:201–220, 2006), predating the oldest known rocks – the ~4.0–4.03 Ga-old Acasta Gneiss (Bowring and Williams Contrib Mineral Petrol 134:3–16, 1999). The zircon data suggest development of low temperature hydrosphere predated the ~3.8 Ga Isua supracrustal belt of southwest Greenland by some 600 Ma. These greenstone belts contain a wealth of primary textural, mineralogical, geochemical and isotopic features allowing detailed insights into the nature of near Earth surface as well as deep seated plutonic processes. Early Archaean (>3.6 Ga) gneisses underlie approximately 10,000 km^2 of Earth surface (Van Kranendonk et al., Paleo-archaean development of a continental nucleus: the east Pilbara terrain of the Pilbara craton. In: Van Kranendonk MJ, Smithies RH, Bennett VC (eds) Earth's oldest rocks. Developments in Precambrian geology 15. Elsevier, Amsterdam, pp 307–337, 2007a, Terra Nova 19:1–38, 2007b). Terrains containing >4.0 Ga (Hadean) detrital and xenocrystic zircons include the Mt Narryer-Jack Hills Terrain (<4404 Ma), Acasta Gneiss (4.03–3.94 Ga) (west Slave Province, northwest Canada) and East Antarctic Gneisses (4.06–3.85 Ga).

Keywords Archaean • Hadean • Zircon • Mount Narryer • Jack hills • Acasta gneiss • Isua belt • Itsaq gneiss

The spatial juxtaposition of the Earth and Moon early in the history of the solar system (Ringwood 1986) requires both have been affected by early meteorite bombardment periods, documented on the Moon at ~3.95–3.85 Ga (Ryder 1990) and earlier periods. However, to date no shock-produced planar deformation features were identified in zircons from the Mt Narryer and other early Archaean terrains. The presence of ~3.85 Ga banded iron formations in southwest Greenland (Appel et al. 1998; Nutman et al. 2010) may signify yet older mafic volcanism with consequent ferrous iron enrichment of the early hydrosphere under unoxidizing conditions of the Archaean atmosphere (Chap. 11). The absence to date in pre-3.75 Ga terrains of un-metamorphosed or low-grade supracrustal rocks retards a search for signature of asteroid impacts. A possibility that terrestrial zircons >4.0 Ga may have been derived from impact-produced felsic and anorthositic melt sheets, contemporaneous with pre-4.0 Ga lunar mega-impacts such as the South Pole Aitken, requires further examination.

The following sections discuss the principal features of high-grade gneisses and relic zircons in geochronological sequence, with implications to the nature of the Hadean (pre-3.8 Ga) crust.

2.1 Mount Narryer >4.0 Ga Detrital Zircons

The 30,000 km^2-large Narryer Terrain, northwestern Yilgarn Craton, Western Australia (Myers 1988; Wilde et al. 2001), consists of granitoids and granitic gneisses ranging in age from early to late Archaean interlayered with banded iron formation, mafic and ultramafic intrusive rocks and metasediments (Williams and Myers 1987; Myers 1988; Kinny et al. 1990). The terrain contains ~3,730 Ma gneiss-hosted zircons and <4,404 Ma detrital zircons hosted by ~3 Ga-old quartzites (Fig. 2.1) (Froude et al. 1983; Wilde et al. 2001; Watson and Harrison 2005; Harrison et al. 2005; Valley et al. 2006; Nemchin et al. 2006; Pidgeon and Nemchin 2006; Cavosie et al. 2004) (Figs. 2.2, 2.3 and 2.4). The gneiss terrain is defined in the west by the Darling Fault, and in the north by the Errabiddy shear zone. Nutman et al. (1993) regards the Narryer Terrain as an allochtonous thrust block over the 3.0–2.92 Ga granites of the Youanmi.

Fig. 2.1 Photographs of quartzite and conglomerate, Jack Hill, northwestern Yilgarn Craton, Western Australia. (**a** and **b**) Jack Hills, Western Australia, site of Hadean zircons as old as 4.4 billion years, the oldest known samples of Earth; (**c**) Outcrop of quartzite pebble-bearing meta conglomerate at Jack Hills. Courtesy John Valley

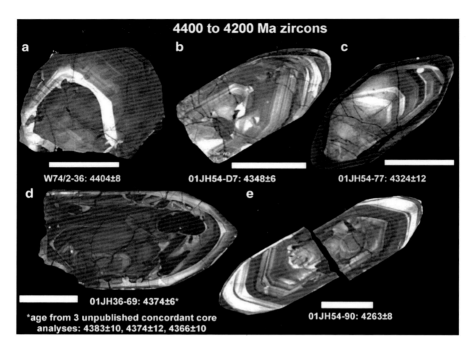

Fig. 2.2 Cathodluminescence images of five 4,400–4,200 Ma detrital zircons from Jack Hills. Details of these grains are presented in Cavosie et al. (2004, 2005, 2006). Ages are in Million years. Uncertainties in Pb-Pb ages are 2 SD. Scale bars = 100 μm (From Cavosie et al. 2007; Elsevier, by permission http://www.sciencedirect.com/science/bookseries/01662635/15)

Hadean to early Archaean ~4.4–3.9 Ga-old ages reported from 251 zircon grains from ~3.0 Ga Jack Hills quartzite sediments indicate a peak magmatic activity during ~4.2–4.0 Ga. Seventeen grains yielded ages >4.3 Ga, mostly from a single arenite layer (Cavosie et al. 2004). A 4,352 Ma-old zircon grain measured from Maynard Hills (Wyche et al. 2004) constitutes the oldest terrestrial age record measured to date. The old zircons may display zoned rims representing magmatic overgrowths, some of which dated as ~3,360 and ~3,690 Ma, consistent with granite ages from the Narryer Terrain, indicating the ~4.3 Ga zircons were re-incorporated in granitoid magmas (Cavosie et al. 2004).

Conglomerates of Jack Hills containing early zircons are dominated by quartzite clasts (Fig. 2.1c). To date no granitoid clasts were encountered, which restricts considerations regarding provenance terrains of the zircons. The U-Pb ages, morphology, zoning and geochemistry the zircon population of the Narryer Terrain suggest derivation from heterogeneous sources (Wilde et al. 2001). About 12 % of the zircons yield ages older than 3.9 Ga (Pidgeon and Nemchin 2006). The bulk of the zircons from Mt Narryer and eastern Jack Hills are consistent with granitic provenance similar to adjacent gneisses in terms of age, oscillatory zoning and U-content, whereas zircons from the western Jack Hills are of different composition and origin (Crowley et al. 2005).

2.1 Mount Narryer >4.0 Ga Detrital Zircons

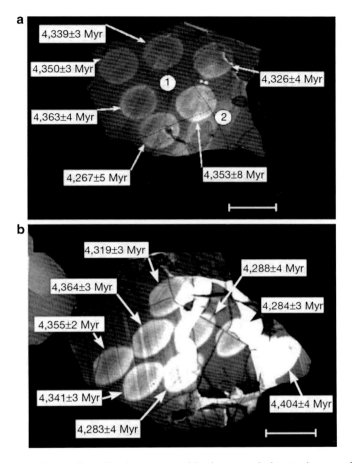

Fig. 2.3 4.4 Ga Zircons Cathodluminescence and back-scattered electron images of a detrital zircon crystal grain W74/2–36 from Jack Hills. Scale bars are 50 µm; (**a**) lighter circular areas are the SHRIMP analytical sites and the values record the ^{207}Pb/^{206}Pb age of each site. The two white areas (1 and 2) represent the approximate location of the δ^{18}O oxygen analytical spots. (**b**) Cathodluminescence image showing the sites of analysis and the ^{207}Pb/206Pb ages for each spot (From Wilde et al. 2001; Nature, by permission)

Palaeo-thermometry studies using Ti levels in zircon and oxygen isotopes support contact with water (Mojzsis et al. 2001; Peck et al. 2001; Wilde et al. 2001; Cavosie et al. 2005; Watson and Harrison 2005). Suggestions of a "cool" early Earth (Valley 2005; Valley et al. 2006) are supported by studies of Ti/zircon of >4.0 Ga-old zircons, inferring temperatures of ~700 °C consistent with modern-day geodynamic conditions (Watson and Harrison 2005). From post ~2.6 Ga, however, a long term rise in the maximum ^{18}O/^{16}O ratios indicates further overall cooling of the Earth crust (Figs. 11.2 and 11.3) (Valley 2008). Preferential retention of Lu relative to Hf in the mantle and the low initial ^{176}Hf/^{177}Hf of most Jack Hills zircons imply derivation

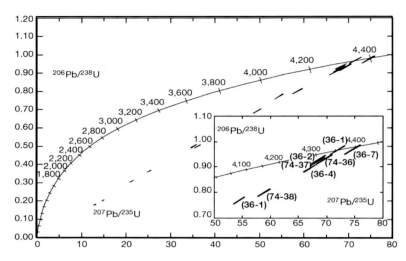

Fig. 2.4 Combined concordia plot for grain W74/2-36 showing the U-Pb results obtained during two analytical sessions. The inset shows the most concordant data points together with their analysis number (From Wilde et al. 2001; Nature, by permission)

by reworking of low Lu/Hf crust. The low ^{176}Hf/^{177}Hf and ^{18}O/^{16}O isotope values of the zircons thus suggest that within ~100 Ma of its formation Earth had settled into a pattern of crust formation, erosion, and sedimentary cycling similar to that produced during younger periods. Many zircons younger than 3.6 Ga have higher initial ^{176}Hf/^{177}Hf ratios, which suggests mantle contributions (Bell et al. 2011).

The assumption of a Hadean continental crust (Harrison et al. 2005) is difficult to reconcile with the likely arrest of the gabbro to eclogite transition under high Archaean and Hadean geotherms, which would have retarded subduction of mafic and production of granitic continental crust (Green 1981). Further, the T_{Zr} values correspond to near-eutectic zirconium saturation temperatures, which could have included equivalents of the Archaean tonalite-trondhjemite-granodiorite (TTG) suite (Chap. 4). Interpretations of zircon populations in terms of the composition of provenance terrains are likely to suffer from a major sampling bias in favor of zircons, as contrasted to a paucity in signatures of labile mafic-ultramafic and other lithic detritus eroded during current reworking. To cite an example, the composition of late Archaean conglomerates, such as in the Moodies Group of the Barberton Greenstone Belt (Chap. 7) or the Lalla Rookh Formation of the Pilbara Craton (Chap. 8), hardly represents the composition of the underlying greenstone-granite terrains.

2.2 Acasta Gneiss and Hadean Zircons

The Acasta Gneiss complex, west Slave Province, northwest Canada, includes relic 4.03–3.94 Ga granitoids and diorites with granitic protoliths indicated as ~4.2 Ga-old (Bowring and Williams 1999; Bleeker and Stern 1997; Iizuka et al.

2.2 Acasta Gneiss and Hadean Zircons

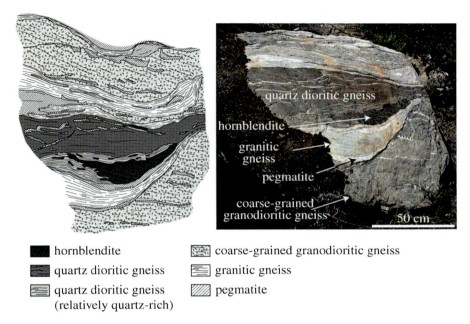

Fig. 2.5 Sketch map and photo of outcrop of quartz dioritic, coarse-grained granodioritic and granitic gneisses, with minor pegmatite and hornblendite layers. The outcrop displays granitic gneiss intruding quartz dioritic and coarse-grained granodiorite gneisses. The quartz diorite gneiss differentiated into relatively quartz-rich quartz diorite gneiss leucosome and residual layers of hornblendite, indicating anatexis of the quartz diorite by intrusion of granitic magmas (From Iizuka et al. 2007; Elsevier, by permission)

2007) (Fig. 2.5). The un-radiogenic Hf isotopic composition and the geochemical and mineralogical properties of the ~4.2 Ga zircon xenocrysts suggest crystallization in granitic magmas derived by reworking of older Hadean felsic crustal rocks (Iizuka et al. 2007). Calculations based on $^{176}Lu/^{177}Hf$ ratios of the ~4.2 Ga zircons and juvenile magma of chondritic Hf isotopic composition allow estimates of mixing ratios and metamorphic reworking, suggesting that about half of the samples have a reworking index higher than 50 % and implying extensive reworking of Hadean granitoid crust (Iizuka et al. 2007). Zircon zonation patterns (Fig. 2.6) include evidence for metamorphic overgrowth of ~3.66 Ga zones around ~3.96 Ga cores, likely associated with intrusion of granites. Younger magmatic events include ~3.74–3.72 Ga granitoids, >3.6–3.59 Ga diorites and granitoids (Iizuka et al. 2007). ~4.2 Ga ages are obtained from model Hf zircon data and ^{40}Ar-^{39}Ar ages of ~4.3 Ga within banded iron formation, also obtained in the Isua supracrustal belt (Kamber et al. 2003) (Sect. 2.5). U-Pb zircon data indicates mafic magmatic activity at ~4.0 and ~3.6 Ga. Evidence for younger magmatic events in the West Acasta terrain includes ~3.36, ~2.9, ~2.6 and ~1.9–1.8 Ga felsic magmatism and related anatexis and metamorphism associated with the Wopmey orogeny

Fig. 2.6 Cathod-luminscence images of zircons: (**a**) from granitic gneiss AC458 with a protolith age of 3.59 Ga, showing that some zircons contain xenocrystic cores; (**b**) from granitic gneiss AC584 with a protolith age of 3.94 Ga, showing the overgrowth of oscillatory-zoned zircon during recrystallization/anatexis at 3.66 Ga, which is coincident with the protolith age of an adjacent tonalitic gneiss. Scale bars are 50 μm. Values record 207Pb/206Pb ages (2σ). "D" and "T" represent analytical spots of LA-ICPMS dating and LA-ICPMS trace element analysis, respectively (Iizuka et al. 2007; Elsevier, by permission)

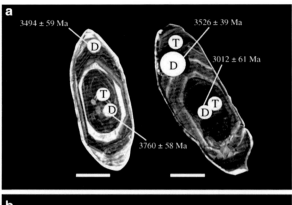

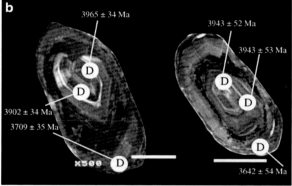

(Hodges et al. 1995; Sano et al. 1999), followed by ~1.26 Ga MacKenzie mafic dyking event (LeCheminant and Heaman 1989; Bleeker and Stern 1997).

2.3 Northeastern Superior Province Gneisses

The Nuvvuagittuq (or Porpoise Cove) greenstone belt, located on the eastern coast of Hudson Bay, Northeastern Superior Province, Canada, dated at 3.8 Ga, is composed of Archaean plutonic suites containing 1–10 km-wide belts of amphibolite and granulite-facies greenstones traced up to 150 km along strike (Percival and Card 1994; Percival et al. 1992; Leclair et al. 2006). The terrain includes one of the world's oldest known mantle-derived assemblages (O'Neill et al. 2007), comprising both Al-depleted and Al-undepleted komatiite volcanics and isotopic positive εNd values indicating long-term preceding mantle-type high $^{147}Sm/^{144}Nd$ ratios and depletion in the light rare earth elements (LREE). U-Pb age determinations define younger 3.8–1.9 Ga domains spatially defined by aeromagnetic anomalies (Leclair et al. 2006). Two principal age terrains are identified, including (A) The east Arnaud

River terrain consisting of <2.88 Ga rocks with Nd depleted-mantle model ages (T_{DM}) <3.0 Ga, and (B) The west Hudson Bay terrain including the Nuvvuagittuq greenstone belt which contain ~3.8 Ga inherited zircons (Stevenson et al. 2006). Mafic components of the gneisses include cummingtonite amphibole with positive Eu and Y anomalies, similar to those of Algoma-type banded iron formations of marine exhalite origin and seawater signature (Fryer 1977; Graf 1978; Fryer et al. 1979). Banded iron formation associated with the greenstone belts display heavy Fe isotopic composition (0.1–0.5‰ amu) relative to those of associated gneisses and amphibolites, indicating origin as marine chemical precipitates, by analogy with heavy Fe isotopic composition of the Akilia banded iron formation, southwest Greenland. The origin of banded iron formations includes interpretations in terms of precipitation of marine exhalations (Jacobsen and Pimentel-Klose 1988; Olivarez and Owen 1991), oxidation of Fe^{+2} derived from deep ocean basins and deposition under anoxic conditions (Holland 2005) and anaerobic microbial activity (Konhausser et al. 2002). The REE and Y profiles and the Fe isotopic compositions of Nuvvuagittuq's BIF confirm their origin as marine exhalites (O'Neill et al. 2007).

2.4 Wyoming Gneisses

The northwestern Wyoming Craton and the Granite Mountain region of central Wyoming include outcrops of 3.4–3.2 Ga rocks (Mueller and Frost 2006) and evidence from Nd and Pb isotopic indices suggesting 3.6–4.0 Ga precursor source terrains (Wooden and Mueller 1988; Frost and Frost 1993; Mueller and Frost 2006; Chamberlain et al. 2003). The Archaean rocks are exposed in the cores of Laramide (Cretaceous) uplifts, bordering Proterozoic terrains and late Archaean ~2.67 Ga terrain of the Teton and Wind River Ranges (O'Neill and Lopez 1985; Mueller et al. 2002; Frost et al. 2006). Other late Archaean terrains occur in the North Snowy Block, southwestern Montana and in south-central Wyoming (Chamberlain et al. 2003).

Detrital zircons from 3.3 to 2.7 Ga quartzite of the eastern Beartooth Mountains, western Beartooth Mountains and Tobacco Root Mountains and Hellroaring plateau are dominated by 3.3–3.4 Ga grains and also include >3.6 Ga zircons, including grains 3.9–4.0 Ga-old (Mueller and Frost 2006). Lu-Hf model ages of 3.3–3.4 Ga from bulk zircon samples in quartzites suggest derivation from a relatively juvenile magmatic event (Stevenson and Patchett 1990). Early Archaean zircons also occur in the northwest part of the Wyoming Craton, which contains extensive pelitic, quartzitic and carbonate metasediments interleaved with gneisses (Mogk et al. 1992, 2004). Zircon xenocrysts from these meta-sediments yield ages of ~3.3 Ga from trondhjemite from the northern Madison Range and up to 3.93 Ga from the Tobacco Root Mountains. Tonalitic orthogneiss with magmatic ages of ~3.3 Ga east of the Granite Mountain contains early Archaean xenocrystic zircon cores mostly yielding ages of about ~3.3 Ga with some grains yielding ages of ~3.42, ~3.48, ~3.59 and ~3.80 Ga (Kruckenberg et al. 2001). Cores of xenocrystic zircons from the southwest Wind River Range include 3.8–3.65 Ga internal domains with low U (34 and 25 ppm) and high Th/U (0.70 and 0.74) values, suggesting crystallization in mafic or high grade metamorphic rocks.

Chamberlain and Mueller interpreted the evidence in terms of the pre-existence of a major ~3.3 Ga craton extending from the Oregon Trail belt in south-central Wyoming to the Great Falls Tectonic zone and Dakotan segment of the Trans-Hudson Orogen. Prevalence of 3.3–3.4 Ga detrital zircons in the eastern Beartooth Mountains suggests proximal sources of terrains of this age. Elevated isotopic $^{207}Pb/^{204}Pb$ ratios in upper Archaean rocks suggest derivation from high-μ Hadean-aged basaltic proto-crust and mantle. Nd isotopic data vary widely, in part suggesting mantle–derivation and in part incorporation of 3.8–3.6 Ga components in the sources of younger magmatic events (Kamber et al. 2003). Chamberlain and Mueller interpreted the evidence in terms of combination of erosion of early Archaean cratons and subduction of basaltic crust including craton-derived sediments into the mantle. Cratonization of the Wyoming Craton about ~3.5, ~3.3 and 3.0–2.85 Ga was associated with multiple intrusions of tonalite-trondhjemite-granodiorite and K-granites in the Beartooth-Bighorn Magmatic Zone, forming broad plutonic blooms rather than arc trench-type linear belts (Frost et al. 2006; Mueller and Frost 2006). Evidence of modern-style lateral accretion and arc magmatism is suggested at 2.67–2.63 Ga in the southern part of the craton. Possible geophysical evidence for buried equivalents of the ~3.3 Ga craton is provided by a, at least 100,000 km^2 large, 15–25 km thick and seismically-fast lower crustal layer imaged beneath the Bighorn subprovince by the Deep Probe experiment (Nelson 1999; Gorman et al. 2002; Chamberlain et al. 2003), imaged north to the Great Falls tectonic zone and much of central and eastern Montana.

2.5 Itsaq Gneisses and Supracrustals, Southwest Greenland

The Itsaq Gneiss Complex contains tonalitic, granitic, minor quartz-dioritic and ferro-gabbro gneisses dated as ~3,870 and 3,620 Ma, associated with amphibolite, chemical sediments, minor felsic volcanics, gabbro-anorthosite suites and slivers of >3,600 Ma peridotite. Intrusive tonalites are compared to the Archaean tonalite-trondhjemite-granodiorite suite in terms of their geochemistry, including Sr and Nd isotopic mantle-derived signatures and are interpreted in terms of partial melting of hydrated basalt converted into eclogite (Nutman et al. 1993, 2007). The northern part of the Itsaq gneiss complex contains the 35 km-long Isua supracrustal belt (Fig. 2.7), originally dated as ~3,700 Ma (Moorbath 1977), 3,800 Ma (Baadsgaard et al. 1984) and 3,710 Ma (Nutman et al. 2007), including meta-volcanic amphibolite and banded iron formations. The primary volcanic and sedimentary textures are mostly overprinted by penetrative deformation, leaving only small low-strain least-deformed structural windows where original features such as pillow lava in amphibolite are observed (Komiya et al. 1999; Nutman et al. 2007). The chemical

2.5 Itsaq Gneisses and Supracrustals, Southwest Greenland

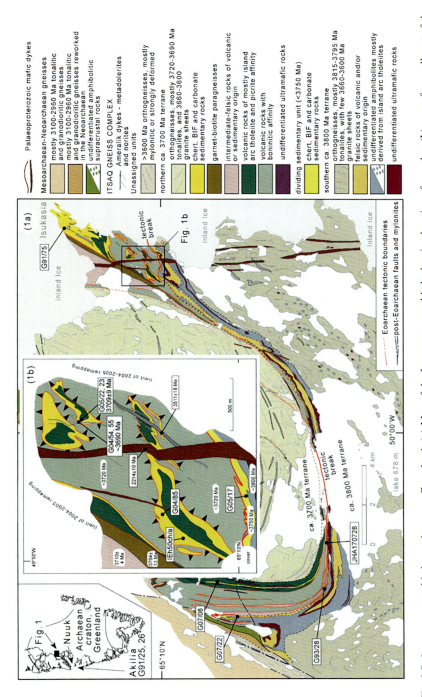

Fig. 2.7 Isua supracrustal belt, southwestern Greenland: (**a**) Map of the Isua supracrustal belt showing location of samples; (**b**) map of a small part of the eastern part of the Isua supracrustal belt, showing the geological relationships between altered intermediate basaltic-andesite(?), pillowed volcanic rocks, chemical sedimentary rocks and an overlying thrust sheet of boninitic amphibolites. The rocks were deformed prior to intrusion of Ameralik dykes ~3,510 Ma (From Nutman et al. 2010; Elsevier, by permission)

Fig. 2.8 Isua supracrustal belt: (**a**) Relict pillows in amphibolite facies ~3.8 Ga meta-basalts of arc-tholeiite to picrite compositions from the southern flank of the Isua Belt; (**b**) Low strain zone in ~3,760 Ma Isua banded iron formations. Note the grading of the layering and the layering on a cm-scale; (**c**) Agglomerate clasts in a ~3.7 Ga Ma felsic intermediate unit from the eastern end of the Isua belt; (**d**) Original layering preserved in quartz + dolomite + biotite +/− hyalophane (Na-Ba feldspar) metasediments, likely derived from marl and evaporites (Courtesy Allen Nutman)

composition of the amphibolites compares with arc basalts (Polat et al. 2002; Jenner et al. 2006). Pillow basalt (Fig. 2.8a) and felsic meta-volcanic schist (Fig. 2.8c) preserving pre-deformation layering and graded volcanic breccia and conglomerate, containing 3,710 Ma zircons (Nutman et al. 1997). Relic banded iron formations (Moorbath et al. 1973; Dymek and Klein 1988) and likely sedimentary calc-silicate are mostly deformed, displaying transposed layering, where local primary sedimentary banding is reserved (Fig. 2.8b–d). The supracrustal relics are tectonically intercalated with the tonalites and are separated by ~3.6 Ga mylonite shear zones.

The Isua metasediments display low Al_2O_3 (<0.5 %) and $TiO2$ (<0.05 %) levels and the Ca-Mg-Fe ratios correspond to those of ferroan dolomite – siderite – Fe-oxide sediments, with the sedimentary origin corroborated by shale-normalized REE + Y profiles showing La, Ce, Eu and Y positive anomalies (Fig. 2.9) which correspond to the signature of sea water (Bolhar et al. 2004). Evidence for early life in Isua sediments and in quartz + tremolite + calcite interstices of pillow lava emerges from Chap. (11):

2.5 Itsaq Gneisses and Supracrustals, Southwest Greenland

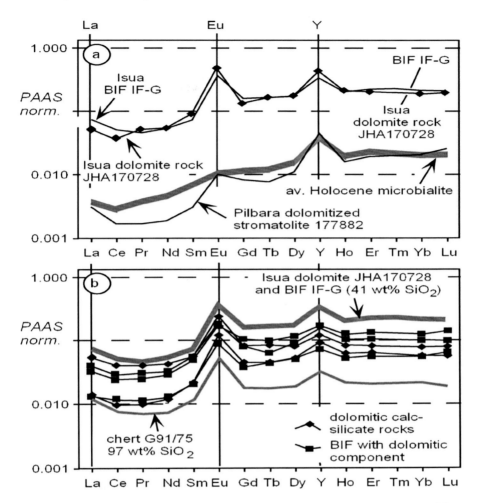

Fig. 2.9 REE+Y patterns normalized to Post-Archaean-Australian-Shale (PAAS). (**a**) IF-G Isua Banded iron formation and metamorphosed dolomite (JHA170728), modern Mg-calcite microbialite from Heron Island, Great Barrier Reef and a Warrawoona ~3.45 Ga low-temperature dolomitized stromatolite; (**b**) Isua dolomitic calc-silicate rocks and BIF with a dolomitic component (From Nutman et al. 2010; Elsevier, by permission)

(A) Experimental evidence for anoxic microbial origin low-temperature dolomites (Nutman et al. 2010; Vasconcelos et al. 1995; Roberts et al. 2004), and
(B) The likely precipitation of banded iron formations either by microbial oxidation of iron or by photolysis by ultraviolet radiation under the ozone-poor or ozone-free Archaean atmosphere (Konhauser et al. 2002).

Granulite facies banded iron formations and calc-silicates observed in the ~3,850 Ma Akilia terrain, southwest Greenland, displaying similar compositions as reported from the Isua supracrustal belt, suggest early traces of microbial life at the end of the late heavy Bombardment defined about 3.95–3.85 Ga (Ryder 1990, 1991, 1997) (Chap. 11).

2.6 Siberian Gneisses

The Aldan Shield of the Siberian Craton north of Lake Baikal is dominated by the Olekma granite–greenstone terrain and gneisses and granitoids of the tonalite-trondhjemite-granodiorite series (TTG). The gneisses yield $T_{Nd}DM$ ages of 3.69–3.49 Ga and are intruded by gabbro at 3.1–3.0 Ga (Rosen and Turkina 2007). The terrain includes mafic-ultramafic and felsic volcanics, dated as 3.2–3.0 Ga, as well as granulite–gneiss suites. The terrains have been interpreted as micro-continents amalgamated during 2.8–2.6 and 2.0–1.7 Ga. Mid-Archaean 3.57–3.51 Ga granitic crust in the Onot and Bulun granite–greenstone domains displays evidence of reworking, recycling, and magma intrusion at 3.41–3.28 Ga. 3.22–3.0 Ga crust in the Sutam granulite–gneiss terrain and 3.4–3.3 Ga of the Sharyzhalgay granulite–gneiss domain display overlap (Rosen and Turkina 2007). The Anabar Shield in the northeast Siberian Craton includes granulite–gneiss terrain dominated by 3.35–3.32 Ga metamorphosed calc-alkaline volcanics, TTG and mafic gneisses, followed by 3.16 Ga enderbites derived by partial melting of mafic source and reworked at 3.1–3.0 Ga. Derived sediments have $T_{Nd}DM$ model ages of 3.19–3.00 Ga (Rosen and Turkina 2007). Granite-greenstone terrains include the Onot and Bulun belts which include TTG plutons with mafic enclaves dated by Sm-Nd ($T_{Nd}DM$) as 3.57 Ga and 3.51 Ga. U-Pb zircon ages of 3.41–3.28 Ga represent metamorphism and contamination. Paleo-pressure and temperature estimates suggest an origin of TTG gneiss under pressures of ~10 kb. The Irkut and Kitoy granulite–gneiss domains contain intermediate and felsic granulites with zircon cores ~3.4–3.3 Ga-old and Nd model ages of 3.4–3.1 Ga. These rocks have undergone high-pressure metamorphism at 2.5–2.4 Ga and granulite facies metamorphism at ~2.0–1.85 Ga. Meta-sedimentary enclaves associated with the gneisses yield model Nd_{DM} ages of 3.6–2.8 Ga. A derivation by erosion of 3.64–3.46 Ga tonalitic gneisses is likely. Diorite gneisses and granites have generally younger ages of 3.2–3.0 Ga. The eastern part of the Aldan shield contains Late Archaean and paleo-Proterozoic rocks.

2.7 North China Craton

The Anshan region of the northern China craton, northeastern China, contains extensive >3.6 Ga gneisses, including some ~3.8 Ga components: (1) Baijiafen mylonitized trondhjemitic rocks; (2) Dongshan banded trondhjemitic rocks, and (3) Dongshan meta-quartz diorite (Song et al. 1996; Liu et al. 1992). Associated younger Archaean rocks include 3.3 Ga Chentaigou supracrustal rocks, ~3.3 Ga Chentaigou granite, ~3.1 Ga Lishan trondhjemite, ~3.0 Ga Donganshan granite, ~3.0 Tiejiashan granite, ~2.5 Ga Anshan Group supracrustal rocks, and ~2.5 Ga Qidashan granite (Liu et al. 1992; Wan et al. 2005). The Archaean blocks are rimmed by paleo-Proterozoic collisional orogenic belt extending through Wutai and Hengshan in Shanxi province to Kaifeng and Lushan in Henan province, separating the North China Craton into an eastern and a western block, the latter including meta-sedimentary rocks (Zhao et al. 2005). Tectono-metamorphic events in the North China Craton are manifest by a concentration of zircon ages at 2.55–2.5 Ga (Shen et al. 2005), constituting about 85 % of the North China Craton. Minor age frequency peaks occur at ~2.7, ~1.85 and ~2.15 Ga. Linear granite-greenstone terrains about ~2.5 Ga in age are widespread, including the western Liaoning province. Volcanic geochemical features correlate with island arc basalts (Wan et al. 2005). ~2.5 Ga greenstones and TTG plutonic rocks were metamorphosed within short periods following their formation (Kröner et al. 1996).

2.8 Antarctic Gneisses

The East Antarctic Shield contains 4.06–3.85 Ga-old high-grade metamorphic blocks and 1,400–910 Ma-old Proterozoic and younger formations which were amalgamated in the Cambrian (Cawood 2005; Harley and Kelly 2007). 3,850–3,400 Ga-old suites are dominated by tonalites, granodiorites and granites in terrains located in Gondwana reconstructions south of India, including the Rayner Province and Prydz Belt, and are superposed and overprinted by younger metamorphic belts. Similar early Archaean events are recorded in the Napier Complex and Rayner Province in Kemp Land where an anatectic/metamorphic event at ca. 3,450–3,420 Ma is locally preserved. Harley and Kelly (2007) suggest that much of the geological terrain covered by the East Antarctic ice cap consists of Archaean crust. Major thermal events have been dated as <2,840 Ma to >2,480 Ma, 2,200–1,700 Ma, 1,400–910 Ma and 600–500 Ma (pan-African Orogeny) (Harley 2003). Originally the Archaean blocks were juxtaposed with Gondwanaland crustal segments including Africa, India and Australia (Tingey 1991; Dalziel 1991; Fitzsimons 2000; Harley 2003).

Chapter 3
Early Archaean Mafic-Ultramafic Crustal Relics

Abstract The long-standing controversy between two schools of thought regarding the precedence in the Archaean of either granitic continental crust or, alternatively, of mafic-ultramafic oceanic-like crust has proven to be an elusive *chicken or the egg* type debate, because: (a) relic supracrustal rocks contain xenocrystic zircons indicative of pre-existing felsic igneous rocks and (b) the oldest gneiss terrains are known to contain mafic enclaves representing yet older supracrustals. The abundance of mafic-ultramafic volcanic xenoliths and enclaves within, and spatially and temporally intertwined relations with Archaean TTG (tonalite-trondhjemite-granoidorite) plutonic suites, as indicated by similar isotopic ages to felsic volcanics within greenstone belts, suggests near-coeval relations. Due to the differential resistance to weathering and erosion processes of zircons, quartz and mafic fragments the occurrence of xenocrystic and detrital zircons up to ~4.4 Ga-old cannot be taken as evidence for a Hadean continental granitic crust. Likewise, the Sm-Nd signatures and entrained zircons in Archaean mafic-ultramafic volcanic sequences, such as the Onverwacht Group in South Africa and Warrawoona Group in northwestern Western Australia, do not necessarily imply an underlying granitic basement since felsic materials could well have been shed from neighboring older granitoid-dominated blocks, such as the Swaziland gneiss complex in South Africa and Mount Narryer terrain in Western Australia. Some mafic-ultramafic volcanic sequences may have formed in rifted zones between early gneiss terrains and may in part overlap the older blocks through unconformities.

Keywords Mafic-ultramafic crust • Greenstone belt • Sial nuclei • Tonalite-trondhjemite-granodiorite

Some of the oldest rock units identified in Archaean low-metamorphic grade granite-greenstone terrains and high-grade gneiss-granulite terrains include supracrustal enclaves within orthogneiss-dominated plutons, occurring either as linear supracrustal belts (greenstone belts) or as discontinuous xenolith swarms (Figs. 3.1 to 3.5). Tectonized boundary zones of batholiths contain foliated gneiss-greenstone intercalations derived by deformation of xenolith-bearing intrusive contacts, a process which involves inter-thrusting and refolding of plutonic and supracrustal units (Fig. 3.4) and tectonic reactivation involving penetrative interleaving of gneiss and foliated greenstone slivers, masking primary granite-greenstone relations. The origin of the xenoliths is betrayed by little-deformed intrusive high-angle contact zones between linear greenstone belts and diapiric tonalite-trondhjemite plutons, displaying progressive magmatic injection, disintegration and assimilation by granitic magma (Figs. 3.1, 3.2 and 3.3).

The distribution patterns of mafic-ultramafic xenoliths in Archaean orthogneiss terrains allow identification of the internal geometry of granitoid batholiths. Principal regional to mesoscopic-scale characteristics of xenolith swarms are outlined in several terrains, including Zimbabwe Craton (Fig. 3.1), southern India (Fig. 3.2) and the Pilbara Craton (Figs. 3.3 and 3.4), displaying transitions along-strike and across-strike between greenstone sequences and xenolith chains, which defines the age of the xenolith. The transitions involve a decrease in the abundance and scale of supracrustal enclaves from synclinal keels (Fig. 3.2a) to outcrop-scale xenoliths (Fig. 3.2b), defining early greenstone belts as mega-xenoliths which outline subsidiary gneiss domes within the batholiths (Figs. 3.1 and 3.5). The common contiguity of xenolith patterns places limits on horizontal movements between individual intra-batholithic domal structures, indicating these domes as least-deformed cratonic islands or nuclei within an otherwise penetratively foliated gneiss-greenstone crust, modelled in Fig. 3.5. The domal architecture suggests magmatic diapirism followed by late-stage solid-state uplift. Exposed high-grade metamorphic sectors along batholith-greenstone boundaries likely represent uplift of deep-seated crustal sectors along reactivated boundaries of batholiths (Fig. 3.4). It is possible some isolated mafic-ultramafic xenoliths represent relic residues of partial melting process which gave rise to felsic magmas.

Examples of early Archaean supracrustal enclaves include the ~3.8–3.7 Ga-old Isua outlier (Fig. 2.7), the ~3.85 Ga Akilia xenolith suite in southwest Greenland (Allaart 1976; McGregor and Mason 1977), the pre-3.6 Ga old Uliak swarm in the Nain Complex, Labrador (Collerson and Bridgwater 1979), Dwalile supracrustal remnants in the ~3.7–3.5 Ga old Ancient Gneiss Complex of Swaziland (Anhaeusser and Robb 1978), pre-3.5 Ga old Sebakwian swarms in the Selukwe and Mashaba areas of Zimbabwe (Stowe 1973) (Fig. 3.1), pre-3.4 Ga old Sargur Group en-claves in the amphibolite to granulite facies gneisses of southern Karnataka (Janardhan et al. 1978) (Fig. 3.2a, b) and greenstone xenoliths and intercalations within Pilbara batholiths in Western Australia (Hickman 1975, 1983; Bettenay et al. 1981) (Figs. 3.3 and 3.4). The commonly deep weathering of Na-rich granitoids of the TTG suite and thus relatively poor exposure as compared to quartz-rich K-rich granitoids (Glikson 1978; Robb and Anhauesser 1983) results in a sampling bias.

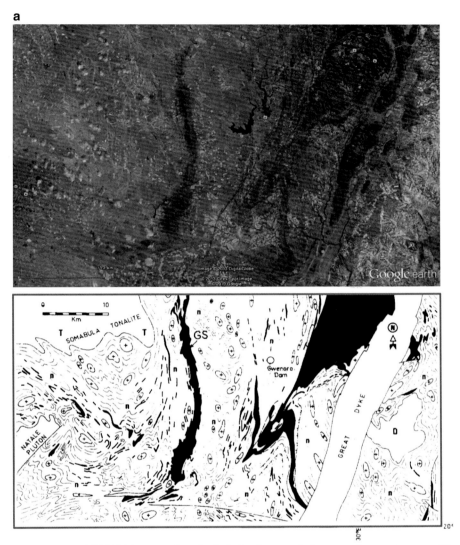

Fig. 3.1 (a) LANDSAT image and geological sketch map of plutons and greenstone xenolith distribution patterns in the Selukwe area. Symbols as for a. SS Selukwe schist belt, GS Ghoko schist belt (Stowe 1973; Wilson 1973; Glikson 1979a; Elsevier by permission); (b) LANDSAT image and geological sketch map of plutons and greenstone xenolith distribution patterns in the Mashaba area, Zimbabwe Craton: *T* tonalite, *D* granodiorite, *A* adamellite, *n* gneiss and migmatite, *p* porphyritic, black areas; greenstone belts and enclaves; trends denote gneissic areas; crosses denote massive plutons (Phaup 1973; Glikson 1979a; Elsevier by permission)

Some of the best documented xenolith patterns have been reported from the Selukwe-Gwenoro Dam area, southern Zimbabwe (Stowe 1973; Taylor et al. 1991), where both across-strike and along-strike transitions occur, for example between

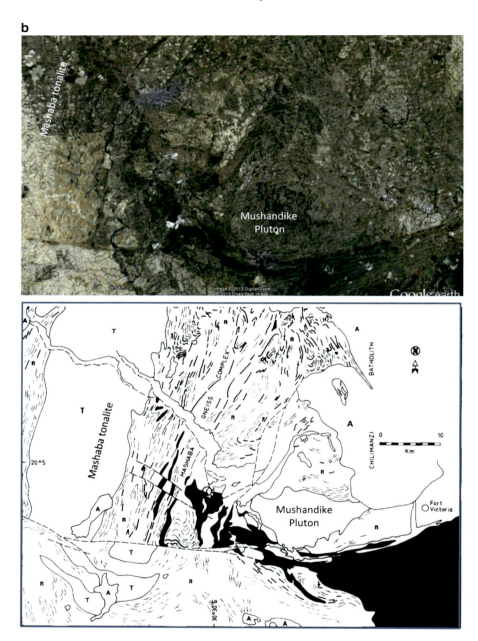

Fig. 3.1 (continued)

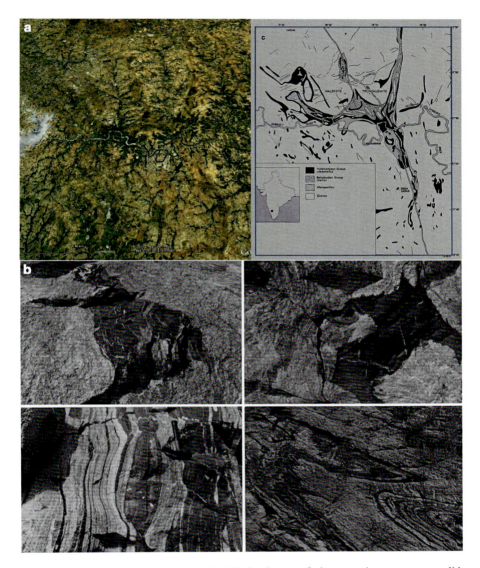

Fig. 3.2 (**a**) LANDSAT image and geological sketch map of plutons and greenstone xenolith distribution patterns in the Holenarsipur area, Dharwar Craton, south India. (**b**) Mafic xenoliths within gneiss, Sargur area, southern Dharwar Craton

the Selukwe and Ghoko schist belt and mafic-ultramafic xenolith swarms (Fig. 3.1a). Here mafic-ultramafic screens outline a ghost stratigraphy contiguous with the greenstone belt with which they merge, as modelled in Fig. 3.5. Whereas magmatic cores within the batholiths are composed of little-foliated TTG, the xenoliths are associated with strongly deformed gneiss and migmatite. Progressive anatexis and

Fig. 3.3 Greenstone xenoliths distribution patterns, Pilbara Craton, Western Australia: (*A*) LANDSAT image of the northern part of the Mount Edgar batholith. *Inset A* relates to image shown in Fig. 3.4a, showing a sheared granite-mafic contact. *Inset B* – a tongue of mafic xenoliths extending from the Talga greenstone belt south into the Mount Edgar batholith; *Inset C* – a tongue of mafic xenoliths extending from the McPhee Reward greenstone belt southeastward into the Mount Edgar batholith. NASA/LANDSAT

digestion of mafic-ultramafic volcanic material was achieved by lit-par-lit injection of near-liquidus tonalitic magma. Metamorphic effects in supracrustal rocks increase from greenschist facies within intact schist belts to amphibolite facies along contact with granite and gneiss. Supracrustal xenoliths may be retrogressed along their margins. Examples of these relations occur in the Gwenoro Dam complex, including the Somabula tonalite, Natale pluton, and Eastern Granodiorite plutons, regarded as the larger intrusive cores of the gneisses. The oldest gneiss phases are typically uniform, fine-grained Na-rich varieties, whereas younger felsic magma fractions are more fractionated, coarser-grained rocks (Stowe 1973).

Xenolith chains extending from mafic-ultramafic sequences of the Onverwacht Group, Barberton Greenstone Belt (BGB), display contiguous transitions from the main greenstone synclinoria into surrounding gneisses, for example between the Stolzburg and Theespruit tonalitic plutons (Viljoen and Viljoen 1969a, b; Robb and Anhauesser 1983) (Fig. 7.2). Anhaeusser and Robb (1978) and Robb and Anhaeusser (1983) applied the xenolith distribution patterns to delineate intra-batholith plutons and cells southwest of the Barberton Mountain Land. Agmatites associated with the xenoliths exhibit intrusive relations, including lit-par-lit injection, discordant

Fig. 3.4 Sheared greenstone belt – gneiss boundaries, central Pilbara Craton, showing intimate intercalations of sheared mafic slivers and xenoliths with granitic gneiss. (**a**) Northern boundary between the Talga greenstone belt and the Muccan batholith (see Fig. 3.3 for locality); (**b**) sheared western margin of the Shaw Batholith, including amphibolite facies gneisses, and the eastern part of the Tambourah greenstone belt. NASA LANDSAT

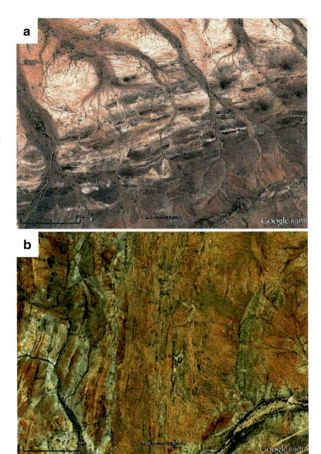

intrusive tongues, mechanical desegregation and progressive assimilation features (Fig. 7.7a). Xenolith fragmentation and digestion patterns depend on the original rock type. Thus felsic tuff units are more heavily veined than mafic rocks, becoming progressively more difficult to distinguish from the intrusive gneiss. Remobilized units composed of partly desegregated and assimilated mafic-ultramafic material are intercalated with gneiss and sheared to obliteration of the original relations between these components (Fig. 7.7a). Small deformed basic dykes display intrusive relations with the gneiss. The occurrence of angular xenolith fragments and assimilated mafic schlieren in close proximity to each other suggests differential viscous flow, where parts of the agmatite froze in situ, while other parts were subjected to continuous flow-deformation and progressive magmatic digestion. Similar features were pointed out by Viljoen and Viljoen (1969a, b) in connection with the Rhodesdale batholith, Zimbabwe, where basal units of the Sebakwian Group, including ultramafic rocks, tholeiitic basalts and fuchsite schist, are identified within gneiss along the periphery of the pluton.

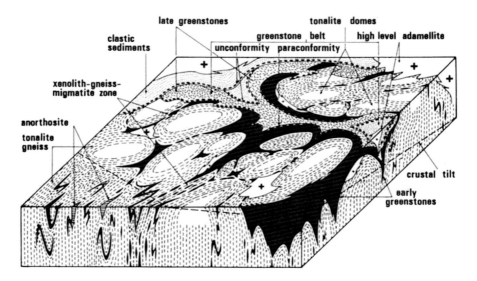

Fig. 3.5 Block diagram illustrating the concept of coeval relations between Archaean granite-greenstone terrains and high-grade terrains. The diagram represents a tilted crustal segment portraying the relations between a major greenstone synclinorium and a high-grade terrain gneiss terrain. Domal tonalites intrude relic ultramafic-mafic volcanic sequences (early greenstones) (*black*). The tonalites become increasingly gneissose (*dashed patterns*) toward their margins and with depth. Late greenstone depositories (*hatched patterns*) and sedimentary sequences (*dotted*) form unconformably and paraconformably above the tonalites and the early greenstones. Loci of maximum late supracrustal deposition coincide with interdomal synclinoria. Deep-seated gneiss root zones of batholiths are characterized by a linear structural grain, attenuated relics of supracrustal enclaves and xenoliths, mafic – ultramafic intrusions and anorthosites (From Glikson 1979a, b; Elsevier by permission)

Hickman (1975, 1983) demonstrated the Mount Edgar batholith, Pilbara Craton, can be divided into at least five oval gneiss domes defined by foliations and intervening amphibolite screens. The preservation of original relationships between supracrustal and plutonic units depends on the primary geometry of their contacts and on subsequent deformation history of the boundary zone. Where supracrustal units strike at high angles to intrusive contacts the original geometry of intrusive contacts may be retained, probably owing to the mechanical strength of interlocked boundaries. Examples of such relations occur along the northern and western flanks of the Mount Edgar batholith, where detached greenstone xenoliths extending from the Talga-Talga Subgroup are engulfed by trondhjemite, attesting to primary intrusive relations (Fig. 3.3). Where the boundary of batholiths and the strike of greenstones are parallel, primary contact relations are mostly overprinted by shear dislocation (Fig. 3.4), representing the susceptibility of these lithological discontinuities to late and post-magmatic reactivation and multiple solid state deformation.

The Shaw batholith of the central Pilbara Craton contains interleaved mafic-ultramafic material, much of it highly deformed. A structural study of the western part of this batholith (Bettenay et al. 1981; Bickle et al. 1983) delineated multiple post-consolidation deformation features affecting tectonically interleaved gneiss-greenstone units. At least three phases of deformation have been identified, including syn-metamorphic horizontal thrusting and recumbent folding. Fold axes and foliations in the gneiss are oriented at low to intermediate angles in relation to the main gneiss-greenstone belt boundary. Upper amphibolite facies metamorphism of some intercalated supracrustal material suggests considerable uplift relative to the juxtaposed Tambourah greenstone belt (Fig. 3.4b). Intrusive gneiss-greenstone relations are recorded. However, the above authors argue against a view of the greenstone intercalations as sheared greenstone-derived xenoliths, pointing out the absence of magmatic assimilation features and suggesting the older gneiss phases are remnants of a granitoid basement predating the greenstones.

According to the early continental basement theory the interlayering of mafic-ultramafic xenoliths and gneiss originated through extensive re-melting of an older granitoid basement. However, anatexis of a pre-existing granitoids basement would have resulted in an abundance of K-rich granitoid neosome fractions, such as are common for example where Nelspruit migmatites and Lochiel Granite are developed by re-melting of Barberton tonalites (Chap. 4) (Fig. 7.2). An alternative view regards the interleaved gneiss-greenstone suite as sheared equivalents of intrusive xenolith-bearing granitoids, for example along the southern border of the Muccan Batholith and southern border of Mount Edgar batholith (Fig. 3.3). These zones range in width from several hundred meters to a couple of kilometers, consisting of gneiss, amphibolite and chlorite schist intercalated with foliated gneiss on a centimeter to meter-scale. Local re-melting resulted in sub-concordant vein systems protruding into the schists. In the Pilbara Craton thermal metamorphism of contact aureoles and xenoliths is mostly up to lower amphibolite facies and rarely attains pyroxene hornfels facies. Late-stage emplacement of Pilbara batholiths was associated with solid-state deformation of solidified upper and marginal sectors, where shearing and retrogression served to obscure original igneous features.

The South Indian Peninsular gneiss complex in the vicinity of Karnataka greenstone belts contains hosts of mafic-ultramafic enclaves (Radhakrishna 1975) (Fig. 3.2a, b). Some of the best examples occur around the Holenarsipur greenstone belt (Naqvi 1981) and the Sargur area (Janardhan et al. 1978). Deformed ultramafic xenoliths and minor pelitic and anorthositic xenoliths of amphibolite facies abound east and west of the Holenarsipur belt. Here supracrustal enclaves form a hierarchy of scales, from a few centimeters to several kilometers in length. Quarry exposures and continuous sections along irrigation canals allow detailed observations of both intrusive and tectonic relationships between gneiss, greenstone and xenoliths (Fig. 3.2b). The density and size distribution frequencies of enclaves in the Hamavati River area are as high as to question any spatial distinctions between greenstone belts and gneiss terrains. Thus, the main Holenarsipur, Nuggihalli and Sandur belts can be regarded as the largest entities within a range of supracrustal enclaves pervaded by tonalitic gneisses on a hierarchy of scales, a relation observed on LANDSAT imagery, on map scale, in individual outcrops and under a thin alluvial veneer in poorly exposed areas.

The south Indian gneiss-greenstone terrains grades into charnockites of the Eastern Ghats metamorphic belt (Fig. 9.5). In the Sargur area hosts of high-grade enclaves of pyroxene granulite, hornblende-rich rocks, ultramafic rocks, quartzite, marble and pelitic schist pervade granulite facies orthogneiss, the ultramafic rocks being volumetrically the most important constituent (Janardhan et al. 1978). The size and continuity of supracrustal enclaves diminish upon transition into the granulite facies terrain, arcuate belt patterns are almost absent, foliated planar structures dominate and original lithologies become difficult to discern. However, syncline keels and supracrustal xenoliths are retained, demonstrating that granulite facies metamorphism was superposed on older granite-greenstone suite. The decreasing abundance of metamorphosed supracrustal material in the high-grade terrain is consistent with the synclinal structure of the supracrustal belts and an increasingly deeper crustal level. Given the northward structural tilt of the Indian Shield (Pichamuthu 1968), variations in abundance, size, and texture of the xenoliths with metamorphic grade allow their use as markers in the study of vertical crustal zonation.

The structure of granite-greenstone terrains and the relations between their components are the subject of long-term controversy. Many regard supracrustal belts as discrete depositories formed in subsiding or down-faulted zones above an older granitoid basement (Baragar and McGlynn 1976; Binns et al. 1976; Archibald et al. 1978; Bettenay et al. 1981; Groves and Batt 1984; Hickman 2012). Evidence for deposition of late Archaean sequences unconformably above older gneisses comes from the Belingwe belt (Bickle et al. 1975), Jones Creek, Yilgarn (Durney 1972), or Finland greenstone belts (Blais et al. 1978). Proponents of an early continental basement theory refer to older cratonic blocks as potential basement for early greenstone sequences. Examples of unresolved relations include: (a) the relations between the <3.55 Ga-old Barberton Greenstone Belt and the ~3.7 Ga-old Ancient Gneiss Complex of Swaziland; (b) the relations between the ~3.0 Ga-old Murchison greenstone belts (Fletcher et al. 1984) and the ~3.7–3.6 Ga-old Narryer Terrain, Northwestern Yilgarn Craton.; (c) the relations between the ~3.8 Ga old Sand River gneisses, Limpopo belt (Barton 1981) and the Murchison greenstone belt, and (d) the relations between the 3.6–3.0 Ga old Western Gneiss Terrain, Yilgarn Block (Gee et al. 1981), and Southern Cross greenstone belts. In each of these instances it is not clear whether the gneiss blocks originally underlay the supracrustals or, alternatively, represent spatially and temporally distinct blocks intervened by rifted greenstone depositories.

A major constraint on an original location of pre-greenstone belts gneiss blocks arises from the chemistry of the tonalite-trondhjemite-Granodiorite (TTG) plutons intruded into the greenstones. Deposition of mid to early Archaean mafic-ultramafic sequences such as the Onverwacht Group and the Warrawoona Group above older gneiss terrains, accompanied or followed by anatexis of such basement, would produce differentiated K-rich granitic magmas, such as occur mainly at late stage of evolution of the greenstone belts. The author's concept of the relation between greenstones, batholiths and their deep level metamorphic analogues is portrayed in Fig. 3.5. Petrological and geochemical features of mafic-ultramafic volcanics are discussed in Sect. 6.1.

Chapter 4
The Tonalite-Trondhjemite-Granodiorite (TTG) Suite and Archaean Island Continents

Abstract Since the 1970s it was realized the bulk of Archaean granitoids intruded into Archaean supracrustal sequences have typically Na-rich composition, comprising tonalite, trondhjemite and granodiorite, the so-called TTG suite (Glikson and Shertaon, Earth Planet Sci Lett 17:227–242, 1972; Hanson, Geochim Cosmochim Acta 39:325–362, 1975; Arth and Hanson, Geochim Cosmochim Acta 39:325–362, 1975; Barker, Trondhjemite: a definition, environment and hypotheses of origin. In: Barker F (ed) Trondhjemites, dacites and related rocks. Elsevier, Amsterdam, pp 1–12, 1979; Glikson, Earth Sci Rev 15:1–73, 1979a; Glikson, J Geol Soc India 20:248–255, 1979b; Moyen et al, TTG plutons of the Barberton granitoid-greenstone terrain, South Africa. In: Van Kranendonk MJ, Smithies RH, Bennett VC (eds) Earth's Oldest Rocks. Developments in Precambrian geology 15. Elsevier, Amsterdam, pp 607–667, 2007). Petrological studies (Green and Ringwood, Contrib Mineral Petrol 18:105–162, 1977), coupled with geochemical and isotopic evidence, demonstrate a derivation of TTG from mafic crustal materials containing isotopic mantle signatures. The evolution of TTG igneous activity is closely correlated with volcanic events within associated greenstone belts. By contrast post-kinematic Archaean granitoids typically comprise quartz monzonites and granites enriched in the large ion lithophile elements (LIL) and display field and petrological evidence for derivation by partial melting of older TTG batholiths. Pre-3.2 Ga TTG tend to occur as oval plutons containing abundant mafic enclaves, with best examples in the Zimbabwe Craton and around the Barberton Greenstone Belt, Kaapvaal Craton, South Africa. By contrast late Archaean granitoids Post-kinematic granitoids are commonly emplaced along boundaries between TTG and supracrustal enclaves, or at high structural levels of batholiths, as is the case with the Lochiel Granite in Swaziland.

Keywords Tonalite • Trondhjemite • Granodiorite • Barberton • Ancient gneiss

4.1 Archaean Batholiths

Granitic terrains have long constituted the terra incognita of Precambrian shields, underlying vast expanses of flat desert, savanna or forest featuring sporadic boulders, monadnocks and low rock benches, which contrast with the commonly better exposed greenstone belts (Figs. 8.1 to 8.4). The largely unknown nature of the granitic terrains was manifested by their representation as undivided pink areas on regional geological maps and their designation in such terms as *'sea of granite'* (Yilgarn, Western Australia) or *'peninsular gneiss'* (southern India) (Pichamuthu 1968; Shackleton 1976). The methodological problems inherent in the study of batholiths have been highlighted by Krauskopf (1970). Glikson and Sheraton (1972) pointed out the dominance of Na-rich tonalite, trondhjemite and granodiorite (TTG) in granite-greenstone terrains and their deep seated equivalents (Viljoen and Viljoen 1969a, b, c; Anhaeusser 1973; McGregor 1973; Arth and Hanson 1975; Tarney et al. 1976; Glikson 1976a, b, c; Hunter et al. 1978; Collerson and Fryer 1978). Mafic and ultramafic xenoliths abound in the TTG (Figs. 3.1 to 3.5, 7.7a). Where the generally poor exposure of these rocks due to weathering of plagioclase resulted in quantitative underestimates, adamellite and K-rich granite are commonly better exposed thanks to their higher quartz content. These rocks tend to occur in three principal modes: (1) as end-members of TTG suites in high crustal levels (Fig. 6.8); (2) as post-tectonic intrusions emplaced along Na-rich granite-greenstone contacts, and (3) as injected bands in trondhjemite—tonalite gneisses (Fig. 7.7b).

Macgregor (1932, 1951) identified two classes of intrusions in the Zimbabwe Craton:

(A) Suess-type batholiths, marked by polydomal 'gregarious' outlines outlined by greenstone belts, enclaves and xenolith screens (Fig. 3.1). In Macgregor's (1932, p. 25) words

> *inclusions of rock of intermediate composition are usually broken up into small masses forming characteristically the darker components of the migmatites, and vary according to the amount of alteration they have undergone from massive amphibolite to pale grey wisps only a little darker than the granite matrix.*

Intrusive contacts of these plutons feature multiple alternations of sheared gneiss and mafic schist whose foliations tend to dip away from the plutonic cores, reflecting their domal structures. Concentric intra-batholithic screens of greenstones are common, representing roof pendants and synclinal keel root zones of greenstone synclinoria (Fig. 3.1). The supracrustals include mafic-ultramafic volcanics and siliceous to ferruginous banded sediments. Progressive assimilation of xenoliths is associated with ptygmatically folded aplite veining controlled by older foliations.

(B) Daly-type batholiths are structurally more uniform and are characterized by abundant coarse-grained and porphyritic rock types, by sharp cross-cutting intrusive contacts, paucity of roof pendants and supracrustal xenoliths, and little evidence of assimilation of xenoliths. This class of intrusions forms massive bodies emplaced into the older Suess-type gneisses and into greenstone belts. They are typically exposed as rugged terrains featuring jointed granitic tors.

The two-fold classification is reflected on Phaup's (1973) map of Zimbabwe granites, where intrusions are classified in terms of petrography (tonalite, granodiorite, adamellite) and structure (gneiss complex, massive granite, porphyritic granite). Oldest components of the gneiss complex include the Tokwe River gneiss (Pb-Pb age: 3475 ± 97 Ma, Taylor et al. 1991) consisting of tonalitic gneiss and migmatite. Massive granites include adamellite, granodiorite and tonalite. Small tonalitic intrusions are emplaced along the margins of, or within, the greenstone belts, examples including the Sesombi, Whitewater, Heany, Bulawayo and Matopos tonalites. Adamellites occur in three principal modes: (1) high-level sheets which dominate the eastern sector of the Zimbabwe Craton, e.g. Zimbabwe, Chilimanzi, Umtali and Salisbury intrusions; (2) small to medium-size intrusions which pierce the older tonalitic gneisses; and (3) small bodies within the greenstone belts. On Phaup's (1973) map the massive granites are commonly separated from greenstone belts by zones of gneiss which represent dynamically deformed margins and/or remnants of older gneisses. A characteristic feature of the older tonalitic gneiss is their wealth in enclaves, screens, bands and schlieren of supracrustal material. In Phaup's (1973, p. 61) words:

There is a close connection in space, time and structural deformation between the gneiss complex and the basement schists, which it almost everywhere bounds and intrudes', and '*inclusions of basement schists of all sizes and in all states of assimilation and dispersion occur nearly everywhere, but are most abundant along the margins and along the root zones of now eroded schist belts.*

Zones of migmatite are common and in many areas are closely associated with greenstone xenoliths, hinting at a relation between migmatization and dehydration of supracrustal material and released water lowering the solidus and viscosity of the granitic magmas. In the Seluke area interbanding and gradations occur between the greenstone belt and supracrustal screens and tonalitic gneiss on a range of scales (Stowe 1973) (Fig. 3.1a). Xenolith-rich gneiss and migmatite zones representing advanced injection of tonalitic magma into root zones and limbs of supracrustal folds are separated by xenolith-poor gneisses which include small oval cores of massive to banded tonalite, designated as 'micro-cratons' (Fig. 3.1a), possibly representing roof cupola of the larger bodies of tonalite (Stowe 1973).

Another example of the Zimbabwe gneiss complex was documented near Mashaba (Wilson 1973) where tonalitic gneisses intruded conformably by the Mashaba Tonalite and Mushandike Granite are in turn succeeded by the high level discordant Chilimanzi batholith (Fig. 3.1b). The gneisses are banded, foliated and migmatitic in structure, containing an abundance of supracrustal mafic remnants, whereas the Mashaba Tonalite consists of homogeneous to weakly marginally banded tonalite. Boundaries between tonalite and the gneiss are gradational. Similar relations pertain between the Mushandike Granite and associated gneisses. Marginal sectors of homogeneous intrusions contain gneiss inclusions derived from the surrounding tonalitic gneisses.

The structural conformity between greenstone belts, xenolith screen-rich gneis--igmatite zones, xenolith-poor gneiss, micro-cratons and larger plutons within Archaean batholiths, and the systematic decrease in abundance of supracrustal material in this order, constitute evidence for progressive migmatite injection and assimilation processes. Supracrustal remnants including amphibolite to granulite facies ultramafi--afic volcanics and banded ironstones may be retrogressed near intrusive margins. Partly digested xenoliths may be enveloped by nebulite and banded gneiss, representing end products of the assimilation process. Whereas early gneiss phases tend to be fine grained, younger bands and apophysae tend to be increasingly coarser grained, porphyritic and pegmatitic.

Similar features pertain to granite-gneiss terrains south and west of the Barberton Greenstone Belt (BGB), Eastern Transvaal (Figs. 6.8 and 7.2). Viljoen and Viljoen (1969a, b) demonstrated a general decrease in the dimension of oval tonalitic domes (*'Ancient Tonalites'*) upwards beneath the Onverwacht anticline, i.e. from the Stolzburg Pluton (~3,460–3,431 Ma) to the Theespruit Pluton (~3,443 Ma) and the Doornhook Pluton. The domes are separated from each other by tongues of komatiitic volcanics of the lower Onverwacht Group (Fig. 7.7a; ~3,548–3,350 Ma, Appendix I). The deformation of the supracrustals is controlled by the configuration of the diapiric tonalites and parallel marginal gneissosity and lineation in the plutonic rocks (Fig. 7.7a). In some instances tonalitic domes developed as parasitic lobes above one another. Two principal varieties of tonalite were distinguished, including leucocratic biotite tonalite (Nelshoogte, Stolzburg and Theespruit plutons) and mesocratic hornblende tonalite (Kaap Valley Pluton) (Viljoen and Viljoen 1969a, b). In both types screens of xenoliths are dominated by komatiite-derived amphibolite and define subsidiary intra-batholitic domal structures southwest of the BGB (Anhaeusser and Robb 1978). The Kaap Valley Pluton (3,229–3,223 Ma) is defined by conspicuous tongues of volcanic rocks branching from the main Barberton greenstone synclinorium, the Nelshoogte belt in the south and the Jamestown belt in the north. These greenstones are more intensely deformed and metamorphosed than central sectors of the BGB (Anhaeusser 1973). The tonalitic diapirs are internally composite, containing multiple injected gneiss, complex banding and veining (Anhaeusser and Robb 1978). The structural concordance and the general decrease in deformation and metamorphic grade from the intrusive margins into central sectors of greenstone synclinoria suggest the dynamic-thermal effects were related to the diapiric intrusive process (Anhaeusser et al. 1969).

Development of the tonalite – greenstone system has been succeeded by emplacement of younger differentiated plutonic bodies which, in contrast to the 'ancient tonalites' invaded stratigraphically higher formations within the greenstone belt, including turbidite-felsic volcanic-BIF association of the Fig Tree Group. The younger granites may occupy synclinal positions in the greenstone belts, e.g. Dalmein Pluton (~3,216 Ma) and Salisbury kop granite (~3,109 Ma). Northwest of the Barberton Greenstone Belt the high-level Nelspruit porphyritic granite (~3.0 Ga) grades in depth into migmatite and gneiss (McCarthy and Robb 1978). To the southeast the high-level Lochiel Granite grades downward into migmatite and tonalitic gneiss which merge and are continuous with the Ancient Tonalites. In Swaziland and southwest of the Barberton Mountain Land the Lochiel granite (~3.0 Ga) and equivalents grade into gneissic terrain, reflecting vertical geochemical and isotopic zonation of the Archaean granitic crust (Viljoen and Viljoen 1969a, b). These relationships are interpreted in terms of anatexis of the tonalites and upward migration of eutectic LIL (Large Ion Lithophile)-enriched partial melts (Glikson 1976a, b, c). A model of vertical stratification of Archaean batholiths is portrayed in Fig. 3.5.

The Pilbara Craton, Western Australia, features oval batholithic structures including the Mount Edgar, Corruna Downs and Shaw batholiths (Figs. 3.3, 6.8, 8.1 to 8.5) (Hickman 1975; Van Kranendonk 2000; Champion and Smithies 1999, 2007). Hickman (1975), who showed Pilbara batholiths consist of subsidiary domes outlined by xenolith-rich screens of greenstones accompanied by sheared gneiss and migmatites, and that these zones form loci of younger post-tectonic granites. Thus the Mount Edgar batholith can be divided into at least five oval bodies (Figs. 3.3 and 8.4). Continuous transitions between the amphibolite screens and greenstone units of the Talga-Talga Subgroup (Fig. 3.3) demonstrate these zones represent deeply eroded keels of greenstone synclinoria. The xenolith screens allow reconstructions of internal batholithic structures, indicating their poly-domal cauliflower-like configuration. Dominant lithologies include trondhjemite, granodiorite and granite, in places truncated by significantly younger felsic plutons. Intrusive relations are observed in relation to greenstones of lower part of the Warrawoona Group (Talga Talga Subgroup: >3,470 Ma, Hamilton et al. 1981; Gruau et al. 1987). Intrusive boundaries are well pronounced where layering orientations of the supracrustals are at high angles to the plutonic bodies (Fig. 3.3), whereas where the strike and the contact are parallel original relations may be obscured by faulting and shearing, reflecting the susceptibility of weak mechanical discontinuities to tectonic dislocation (Fig. 3.4). Parallel to subparallel granite-greenstone boundaries feature multiple alternations of sheared granitic gneiss and amphibolite schist, owing to lit-par-lit injection of gneiss and stress-induced interleaving. Aplite and pegmatite may abound in contact zones. Xenolith abundances and the extent of foliation decrease from the margins of pluton inwards, and cores of the plutons consist of weakly banded to homogeneous granite and gneiss. In places, subsidiary lobes of batholiths protrude into the adjacent greenstones, for example along the northern margin of the Mount Edgar batholith (Figs. 3.3 and 8.4).

Plutonic TTG batholiths in the Yilgarn Craton, southwestern Australia, and the Superior Province, Canada, are mostly more linear than those in the Zimbabwe, Kaapvaal and Pilbara cratons, i.e. NNW tectonic trends in the Yilgarn Craton, EW tectonic trends in the Superior Craton and EW trends in the western Pilbara terrain. Poor outcrop and intense faulting combine to obscure original contact features. Where the development of linear features in upper to late Archaean terrains may signify an onset of plate tectonics (Van Kranendonk 2011), as observed by Hamilton (2003) ophiolites and melange wedges of the circum-Pacific type are missing in Archaean terrains (Chap. 12).

The multi-domal geometry of the batholiths (Fig. 3.5) suggests their evolution by long-term accretion of progressively ascending magmatic increments involving diachronous rise of superheated magma into early mafic-ultramafic crust, assimilating vast volumes of the latter. A likely superheated state of TTG magmas and suppression of the solidus by volatiles derived from the digested supracrustals xenoliths, lowering the viscosity, account for enhanced mobility of the magmas. Concomitant chilling of marginal concentric zones and further upward movements of viscous crystal mush resulted in deformation of solidified margins and of xenolith screens, serving as long-term loci for anatectic events. Post-magmatic isostatic adjustments between low-density felsic plutonic complexes and high-density greenstone synclinoria resulted in intermittent subsidence of the latter, as represented in the Pilbara Craton by the preferential occurrence of lower Proterozoic volcanic outliers of the Fortescue Group above the greenstone synclinoria (Hallberg and Glikson 1980). Petrological and geochemical features of the TTG and post-kinematic granites are discussed in Sect. 6.2.

4.2 Vertical Crustal Zonation

The relations between supracrustal low-metamorphic grade greenstone belts, gneiss terrains (Chap. 2) and root zones of Archaean batholiths are critical for a resolution of the structure and evolution of the Archaean crust. A commonly held view of high-grade metamorphic terrains regards them as spatially and temporally distinct from as well as older than supracrustal greenstone belts (Windley and Bridgwater 1971; Baragar and McGlynn 1976; Windley and Smith 1976; Bridgwater and Collerson 1976, 1977; Moorbath 1977; Young 1978). However, continuous transitions between remnant greenstone belts and gneiss terrain occur in the Zimbabwe Craton (Chap. 3; Fig. 3.1), southwestern Greenland (Sect. 2.5) and southern India (Figs. 3.2a, 9.5), In southern India, thanks to the northward tilt of the Dharwar Craton, a near-continuous gradation along strike occurs between greenstone belts (Dharwar, Nilgiris) and gneiss-granulite complexes (Pichamuthu 1968; Shackleton 1976; Naqvi et al. 1978) (Figs. 3.2a and 9.5). In Manitoba the high-grade Pikwitonei belt represents deep-seated equivalents of the Cross Lake granite-greenstone terrain, western Superior Province (Bell 1971; Ermanovics and Davidson 1976; Weber and Scoates 1978). In the Yilgarn Craton, Western Australia, an eastward crustal is indicated by seismic data along the Perth-Kalgoorlie cross section (Mathur 1976), corroborated by an overall eastward decrease in metamorphic grade (Glikson and Lambert 1976). However, in several of these terrains associated high-grade metamorphic terrains and evidence from xenocrystic and detrital zircons indicate pre-existence of terrains older than the low metamorphic grade supracrustals (Chap. 9). The compositional relations between greenstone belts and sedimentary remnants in high-grade metamorphic terrains provide clues for the original relations between these terrains. High-grade meta-sediments in southwestern Greenland (Isua enclave, Malene Supracrustals) and eastern Labrador (Upernavik Supracrustals) contain K-rich potassic pelites or tuffs, quartzite, carbonate and banded iron formations, referred to by Windley and Smith (1976) as evidence for a distinct tectonic environment. Whereas sediments intercalated with ultramafic-mafic sequences of Archaean greenstone belts are dominated by chert, some greenstone belts contain cross-bedded quartzite such as occur at basal stratigraphic levels of the Dharwar greenstone belts (Srinivasan and Sreenivas 1972) and in the

4.2 Vertical Crustal Zonation

Wheat Belt of southwestern Australia. Minor carbonate sediments occur in the Warrawoona Group of the Pilbara Craton. K-rich tufts and derived sediments occur in the Theespruit Formation, Onverwacht Group. However, many occurrences of 'quartzite' and pelitic sediments in Archaean terrains correspond to recrystallized chert and metamorphosed tuff, respectively. A question arises whether some of the older terrains constituted basement for supracrustal greenstone belts or, alternatively, formed spatially and temporally discrete blocks separating down faulted supracrustal rift zones. In the absence of Paleomagnetic constraints on the position of Archaean terrains, isotopic age determinations offer temporal, if not a spatial, clues in this regard.

Chapter 5
Isotopic Temporal Trends of Early Crustal Evolution

Abstract The application of modern isotopic and geochemical analytical techniques to the study of Early Precambrian magmatic, sedimentary and metamorphic rocks and minerals allows the tracing of temporal isotopic and geochemical trends and the temporal and spatial evolution of Archaean systems. The fractionation of parent and daughter isotopic pairs, such as U-Th-Pb, Sm-Nd, Rb-Sr, Lu-Hf, Re-Os between source mantle, partial melts and crustal rocks allows (a) geochronological age determinations and (b) information regarding the degree of fractionation of precursor materials of analyzed rock suites represented by initial isotopic ratios, i.e. initial $^{238}U/^{204}Pb$, $^{87}Sr/^{86}Sr$, $^{143}Nd/^{144}Nd$, $^{176}Hf/^{177}Hf$, $^{187}Os/^{188}Os$. Distinctions between mantle type precursor materials and crustal precursor materials allow an insight into the source of magmatic formations and their detrital derivatives. For example the low initial $^{176}Hf/^{177}Hf$ ratios of some early zircons hint at Lu-depleted precursor felsic igneous rocks whereas high initial $^{176}Hf/^{177}Hf$ ratios suggest mantle origin, constituting a tool in identification of the pre-history of crustal materials.

Keywords Isotopes • Precursor crust • U-Th-Pb • Rb-Sr • Sm-Nd • Lu-Hf • Re-Os

5.1 U-Th-Pb Isotopes

The differential melt/mineral/parent rock partitioning coefficients of U, Pb and Th, their fractionation between mantle and crust, combined with knowledge of isotopic parent-daughter decay constants, allow insights into the nature of precursors and age of geological materials. Models of evolution of U-Pb-Th isotopic systems (Tatsumoto 1978; Hofmann 1988; McCulloch and Bennett 1994) (Fig. 5.1) are complicated by changes in the composition of the mantle and in variations in the U, Th and Pb mantle/melt partition coefficients with time. The decay constants of

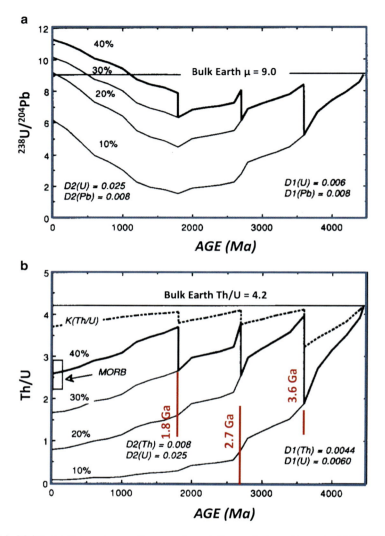

Fig. 5.1 Major mantle-crust recycling events and the evolution of mantle ^{238}U/^{204}Pb ratios (μ values); (**a**) Calculated evolution of U/Pb in average MORB source. The *solid line* shows the evolution where the volume of depleted mantle increases in a stepwise manner from 10 to 40 %. At 1.8 Ga there is a switch in the incompatibility of uranium relative to lead ($D_U > D_{Pb}$) (D = partitioning coefficient) resulting in the post-Archaean depleted mantle undergoing a rapid increase in U/Pb. Modelling assumes a primitive mantle at 4,450 Ma with ^{238}U/^{204}Pb = 9.0. From 4,450 Ma to 4,550 Ma. ^{238}U/^{204}Pb = 1.0, with an initial lead composition at 4,550 Ma of Canyon Diablo; (**b**) Plot of the calculated Th/U ratios in the MORB source vs. time. The Th/U ratio of the mantle decreases continuously as $D_{Th} < D_U$. The sharp steps mark the increase in the volume of the depleted upper mantle from 10 % to 40 %. The present-day relatively low Th/U ratio of MORBs of ~2.5 as inferred from ^{234}U/^{230}Th disequilibrium studies is reproduced in this model. The time average Th/U or K_{Pb} value calculated from the ^{207}Pb/^{206}Pb ratios (*heavy dashed line*) is buffered by the growing volume of depleted mantle and has a present-day value of 3.7, also in good with observations. In our model there is no requirement for an extremely short residence time for thorium in the upper mantle. Calculations assume a bulk Earth or primitive mantle Th/U ratio of 4.2 (From McCulloch and Bennett 1994; Elsevier by permission)

uranium isotopes to lead isotopes allows calculation of the equilibration ages as below:

$$^{235}U\left(e^{\lambda_1 t} - 1\right) = {}^{207}Pb; \text{decay constant } {}^{235}U = 9.849 \times 10^{-10} \text{ year}^{-1};$$
$$\text{half-life } 7.07 \times 10^8 \text{ years}$$

$$^{238}U\left(e^{\lambda_2 t} - 1\right) = {}^{206}Pb; \text{decay constant } {}^{238}U = 1.551 \times 10^{-10} \text{ year}^{-1};$$
$$\text{half-life } 4.47 \times 10^9 \text{ years}$$

Which allows calculation of the age t from plots of $^{207}Pb/^{204}Pb$ vs $^{206}Pb/^{204}Pb$.

The scatter of $^{207}Pb/^{204}Pb$ vs $^{206}Pb/^{204}Pb$ ratios of modern mid-ocean ridge basalts (MORB) delineates an array roughly corresponding to an age of ~1.8 Ga. Mantle evolution models for the period ~4.45 Ga ~1.8 Ga, using element partitioning values (D = mantle/melt) ($D_{Pb}=0.008$; $D_u=0.006$; $D_{Th}=0.0044$), indicate continuous relative depletion of the mantle in Uranium and thus a decline in the U/Pb ratio (µ value) from 9.0 to 6.1. Likewise a decline is observed in the mantle Th/Pb and Th/U ratios during the Archaean and the early Proterozoic, reflecting a preferential extraction of Th relative to Pb and U from the upper mantle into the crust (McCulloch and Bennett 1994) (Fig. 5.1). A modelled mantle-crust mixing event about ~1.8 Ga leads to re-introduction of U to the mantle and an increase in the µ value of the mantle from 6.1 to 11.2. Other mantle-crust interaction episodes are defined at ~3.6–3.5 Ga and 2.7–2.6 Ga, consistent with geological evidence (Chaps. 7 and 8). These changes can be modelled in terms of mixed radiogenic and un-radiogenic components, an increase in the volume of depleted mantle and its consequent mixing with compositionally more primitive lower mantle about ~1.8 Ga (McCulloch and Bennett 1994). The Th/U ratio of the mantle decreases as $D_{Th} < D_U$, represented by low Th/U of MORB (Allegre et al. 1986). A Plot of the calculated evolution of the U/Pb and Th/U in the average mantle source of MORB is related to a step-wide increase in the volume of U-depleted mantle with time from 10 to 40 %, as shown in Fig. 5.1 (McCulloch and Bennett 1994), assuming a primitive mantle with $^{238}U/^{204}Pb$ of 9.0 at ~4,450 Ma.

5.2 Rb-Sr Isotopes

The fractionation of Rb/Sr during partial melting, the incompatibility of Rb relative to Sr vis-à-vis residual phases (pyroxene, plagioclase), and the growth of $^{87}Sr/^{86}Sr$ through the decay of ^{87}Rb, provide parameters for differential isotopic evolution in the mantle and crust, formulated by:

$$^{87}Sr/^{86}Sr_t = {}^{87}Sr/^{86}Sr_{initial} + {}^{87}Rb\left(e^{\lambda t} - 1\right); \lambda = 1.42 \times 10^{-11} \text{ year};$$
$$^{87}Rb \text{ half-life } 48.8 \times 10^9 \text{ years.}$$

Directly measured $^{87}Sr/^{86}Sr$ values of Archaean volcanic rocks (McCulloch and Bennett 1994) tend to be higher than model calculations of the $^{87}Sr/^{86}Sr$ evolution of the mantle which assume an original basaltic achondrite best initial composition (BABI: $^{87}Sr/^{86}Sr = 0.69897$), determined by Papanastassiou and Wasserburg (1969) (Fig. 5.2a). The discrepancy is resolved by assuming higher initial Sr isotopic composition for the Earth at −4450 Ma of $^{87}Sr/^{86}Sr = 0.6995$, defined as BEBI (Bulk Earth Best Initial), as well as lower Rb/Sr partitioning coefficient (D) of 0.006, instead of 0.010 (McCulloch and Bennett 1994), implying a slower rate of growth of $^{87}Sr/^{86}Sr$ in the mantle with time. Other factors include possible variations in the D parameter due to the water contents of melts and thus melt/crystal equilibria related to changes in plate tectonic processes with time (McCulloch and Gamble 1991). Secondary alteration of the mobile Rb and Sr during metamorphism, hydrothermal and weathering processes complicate interpretation of Sr isotopic compositions. This includes uncertainties in application of the Sr isotopic composition of relic Archaean pyroxenes and low Rb/Sr barite deposits to early mantle compositions.

A clear distinction is observed between the range of $^{87}Sr/^{86}Sr$ values of Archaean and post-Archaean igneous and sedimentary rocks, where the Archaean rocks spans $^{87}Sr/^{86}Sr$ values of 0.700–0.705 and Proterozoic rocks up to approximately 0.720 and higher (Fig. 5.2c). This suggests lower ^{87}Rb and $^{87}Sr/^{86}Sr$ values of Archaean precursors materials as compared to mixed crustal and mantle sources of Proterozoic and younger rocks. Flament et al. (2013) discuss the implications of the evolution of the $^{87}Sr/^{86}Sr$ of marine carbonates to continental growth, suggesting the late Archaean rise in $^{87}Sr/^{86}Sr$ of marine carbonates signifies the emergence of the continents rather than enhanced rate of production of continental crust.

5.3 The Lu-Hf System

Fractionation of the elements Lu and Hf during mantle melting processes, where Hf is preferentially incorporated in partial melts, coupled with the decay of ^{176}Lu isotope to ^{176}Hf isotope (half-life of ~35.10^9 years), allows isotopic dating and identification of precursors of crustal rocks (Kinny and Maas 2003; Griffin et al. 2003). The enrichment of the rare earth elements (REE) in zircon renders the Lu-Hf method suitable for determination of precursors of zircons. Isotopic dating using Hf isotopes is based on the relation:

$$\left(^{176}Hf/^{177}Hf\right)_t = \left(^{176}Hf/^{177}Hf\right)_{initial} + \left(^{176}Lu/^{177}Hf\right)_t \cdot \left(e^{t\lambda} - 1\right)$$

(t – Elapsed time; λ – ^{176}Lu β decay constant, commonly held at $1.94.10^{11}$ year^{-1} and revised to $1.86.10^{11}$ year^{-1} (Scherer et al. 2001); ^{177}Hf is used for normalization of Hf and Lu abundances).

Studies of Lu-Hf in zircon use the more accurate U-Pb systematics for definition of the ages of $^{176}Hf/^{177}Hf$ ratios in internal components of the zircon. Departures of

5.3 The Lu-Hf System

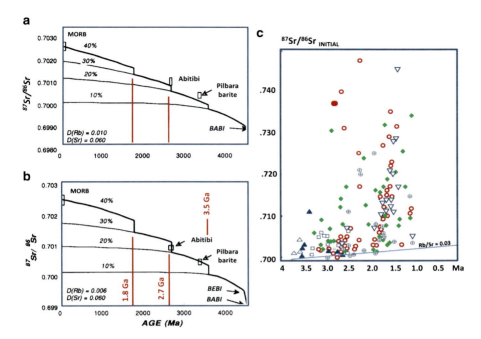

Fig. 5.2 Major mantle-crust recycling events (represented by *red lines*) and the evolution of mantle $^{87}Sr/^{86}Sr$ ratios. (**a**) Plot of the evolution of $^{87}Sr/^{86}Sr$ ratios in progressively growing depleted mantle. The fraction of mantle involved in the formation of continental crust is assumed to increase with time from 10 % to 40 % (by mass) of the total mantle. In this figure the accretion of the Earth is assumed to be essentially complete by 4,550 Ma with the Earth having an initial $^{87}Sr/^{86}Sr$ ratio of BABI (Basaltic Achondrite Best Initial) = 0.69897 and constant D (partitioning coefficient) values for Rb and Sr. The few measurements of the early Archaean mantle represented by data from the Abitibi, Onverwacht and Pilbara greenstone belts are all systematically higher than those predicted by the strontium isotopic evolution model. This may reflect incorporation of crustal-derived radiogenic strontium in the samples, or inappropriate model parameters; (**b**) Plot of the evolution of $^{87}Sr/^{86}Sr$ ratios in progressively growing depleted mantle using modified model parameters. In this figure the accretion of the Earth takes ~100 Ma to complete, and as a result has a higher initial strontium ratio of BEBI (Bulk Earth Best Initial) = 0.69950 at 4,450 Ma. With a higher initial strontium isotopic ratio for the Earth, it is possible to account for the observed $^{87}Sr/^{86}Sr$ ratios in both the Archaean and present-day MORBs (From McCulloch and Bennett 1994; Elsevier, by permission); (**c**) Plots of Rb – Sr isochron ages against corresponding initial 87Sr/86Sr ratios, including data from the North Atlantic Craton, Canada, Australia, Zimbabwe and Transvaal-Swaziland. *Open triangles* – North Atlantic Craton; *circled crosses* – Canadian Shield; *Open circles* – Australian granites; *diamonds* – Australian gneisses; *inverted triangles* – Australian felsic volcanics; *solid triangles* – Zimbabwe plutons; *solid diamonds* – eastern Kaapvaal Craton (From Glikson 1979a; Elsevier by permission)

Hf isotopic composition from chondritic isotopic composition at elapsed time (t) are given by:

$$\varepsilon_{Hf} = \left[\left(^{176}Hf/^{177}Hf \right)_t / \left(^{176}Hf/^{177}Hf \right)_{chondrites} - 1 \right] \times 10^4$$

Positive values reflect relatively high $^{176}Lu/^{177}Hf$ sources and negative εHf values reflect relatively low $^{176}Lu/^{177}Hf$ source, allowing estimates of the relative abundance of mantle vs crustal source compositions. Progressive modification of Lu/Hf ratios upon mantle melting and crustal fractionation raises the mantle $^{176}Lu/^{177}Hf$ to levels higher than those of chondrites and lowers the crustal $176Lu/^{177}Hf$ and $^{176}Hf/^{177}Hf$ ratios (Fig. 5.3).

The Lu/Hf isotopic systematics of Hadean and Archaean zircons indicate a strong early-stage fractionation of the mantle, producing Lu-rich mantle residue and Lu-depleted melts, represented by low εHf (176Hf/177Hf) ratio inherited by the zircons (Kamber 2007) (Fig. 5.3a). ~4.4–4.0 Ga zircons display low εHf values (~2.799–2.802) whereas post ~3.8 Ga zircons display higher εHf values, representing re-enrichment of magma sources in Lu. Kamber (2007) notes a rehomogenisation of 176Hf/177Hf ratios between ~3.96 and ~3.76 Ga, reducing Hf isotopic variability by about two thirds. Comparisons of εHf$_{initial}$ data for the Limpopo belt and worldwide versus isotopic ages indicates initial Lu/Hf ratios lower than those of chondrite and in some instances lower than those of depleted mantle (Zeh et al. 2009, 2014) (Fig. 5.3b). The prevalence of granites with εHf$_{init}$ values below values of CHUR (chondritic upper mantle reservoir) and below DM (depleted mantle) indicates derivation of the bulk of the magmas from low Lu/Hf crustal rocks. It is possible the re-enrichment of crustal materials and zircon in Lu took place in connection with the heavy bombardment (LHB: ~3.95–3.85 Ga – Ryder 1990, 1991) due to impact remelting of residual Lu-rich mantle.

5.4 The Sm-Nd System

The long half-life of 10^9 years for ^{147}Sm decay to ^{143}Nd renders the method useful in isotopic studies of meteoritic and early terrestrial rocks (DePaolo 1983; Dickin 1995). Differential partitioning of Sm and Nd into minerals helps with multi-mineral isochron age studies using phases such as garnet and pyroxene.

The decay of ^{147}Sm to ^{143}Nd by α-decay

$$\left(^{147}Sm \rightarrow {}^{143}Nd \ T_{1/2} = 10^6 \, Ga; \ \lambda = 6.54 \times 10^{-12} \, year^{-1}\right)$$

Fig. 5.3 (continued) and the continental crust (points plotting *below* BSE). Combined data imply an event of rehomogenisation sometime between 0.6 and 0.8 billion years after Earth accretion, here schematically outlined with a *grey polygon*, in which Hf isotope variability was reduced by 2/3 at 0.6 Ga. Irrespective of age and models, note that on average, zircon has unradiogenic Hf plotting well below BSE, indicating that the host melt originated by re-melting of pre-existing crust. Very few zircons reflect direct derivation from mantle (From Kamber 2007; Elsevier, by permission). (**b**) Comparison of combined U–Pb–Hf isotope datasets from the Limpopo Belt with datasets from worldwide sources. Arrows define the evolution of Hadean TTG and mafic rocks. The prevalence of granites with εHf$_{init}$ values below values of CHUR (chondritic upper mantle reservoir) and below DM (depleted mantle) indicates derivation of the bulk of the magmas from low Lu/Hf crustal rocks (From Zeh et al. 2009, 2014; Elsevier, by permission)

5.4 The Sm-Nd System

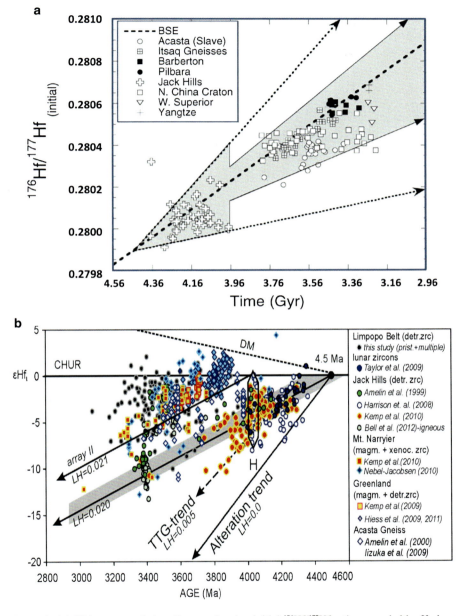

Fig. 5.3 (**a**) Hf isotope evolution diagram showing initial ^{176}Hf/^{177}Hf ratios recorded by Hadean and Paleo-Archaean zircon as a function of time calculated from U/Pb zircon crystallisation age. Position of bulk silicate Earth (*BSE*) reflects the ^{176}Lu decay constant (Scherer et al. 2001). Note that Hadean zircons (pre-4.0 Ga) define a wide spread in Hf isotope composition. Trajectories that encompass this spread (shown as *stippled arrows*) extrapolate to much greater Hf-isotope variability than what is actually observed in younger zircon; namely, Paleoarchaean zircons require trajectories with much less divergence between the depleted reservoir (points plotting *above* BSE)

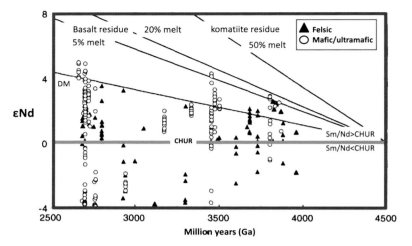

Fig. 5.4 Plot of model T_{Nd} vs. time showing the effects of extraction and storage of basaltic and komatiitic crust on the neodymium isotopic evolution in the residual depleted mantle. Neodymium evolution curves are shown for 100 % storage of tholeiitic crust produced by 5 and 20 % partial melting and komatiite produced by 50 % partial melting also shown are the initial neodymium isotopic compositions for selected Archaean rock (From McCulloch and Bennett 1994; Elsevier by permission)

Is expressed by the relations:

$$^{143}Nd/^{144}Nd = \left(^{143}Nd/^{144}Nd\right)_o + \left(^{147}Sm/^{144}Nd\left(e^{t\lambda}-1\right)\right)$$

The CHUR (chondritic unfractionated reservoir) model, representing material formed in the primitive solar nebula, defines the relations between the Nd isotopic composition of samples relative to the original (chondritic) $^{143}Nd/^{144}Nd$ ratio in the terms:

$$\varepsilon_{Nd,CHUR}(t) = \left[\left(^{143}Nd/^{144}Nd\right)_{sample}/\left(^{143}Nd/^{144}Nd\right)_{CHUR-1}\right] \times 10^4$$

$$^{143}Nd/^{144}Nd_{CHUR} = 0512638$$

Model Sm/Nd ages can be estimated from a single pair of measurements by (1) calculating an assumed growth rate of $(^{143}Nd/^{144}Nd)_{sample}$ from the $(^{147}Sm/^{144}Nd)_{sample}$; (2) assuming the Sm/Nd of the source mantle in terms of either an LREE-depleted mantle or an undepleted chondritic mantle; (Fig. 5.4); (3) plotting the intersection of the sample growth trend with the depleted or chondritic mantle growth trend.

Due to preferential retention of Sm and thus $^{147}Sm/^{144}Nd$ (εNd) in the mantle and the loss of the relatively incompatible Nd to magmas and the crust, the mantle evolved a high εNd, represented in its partial melting products with time ($\varepsilon_{Nd, CHUR} > 1$) vs intra-crustal material ($\varepsilon_{Nd, CHUR} < 1$) (Fig. 5.4). Complications occur: in the Barberton Greenstone Belt time-integrated $Sm_N/Nd_N > 1.0$ indicated by isotopic studies are contrasted with directly analyzed REE concentrations indicating $Sm_N/Nd_N < 1.0$ of the samples (Glikson and Jahn 1985). This suggests secondary LREE enrichment of the mantle prior to partial melting and/or secondary LREE enrichment of the basalts. In the ~3.8 Ga Itsaq gneiss and Isua Greenstone Belt sediments and amphibolite, southwest Greenland, display higher $^{142}Nd/^{144}Nd$ than younger Archaean and modern mafic rocks, suggesting high $^{147}Sm/^{142}Nd$ ratios in their mantle source and thereby early loss of the relatively incompatible Nd from the mantle (Kamber 2007). This suggests extraction of LREE-enriched felsic magmas and of granitic crust from the mantle preceded 3.8 Ga. Thus both εHf and εNd parameters manifest loss of the relatively lithophile Hf and Nd from the mantle at early stages of crustal evolution.

McCulloch and Bennett (1994) developed a 3-component geochemical model including primitive mantle, depleted mantle and crust reservoirs, exploring the relations between mantle depletion and crustal growth with time. A step-wise episodic evolution is suggested by high ε_{Nd} indices representing mantle fractionation and rapid crustal growth at ~3.6, ~2.7 and ~1.8 Ga. The mostly positive ε_{Nd} composition of ~3.8–3.4 Ga Archaean rocks suggest mantle sources have been strongly depleted in Nd and other light rare earth elements (LREE) at ~3.8 Ga (Frey et al. 2004; O'Neill et al. 2007). A wider spread of positive and negative εNd values at 2.7–2.6 Ga indicates a combination of melting of high Sm/Nd mantle source and low Sm/Nd crustal rocks. In McCulloch and Bennett's (1994) model ~10 % of the mantle mass (upper 220 km) is extracted during ~4.5–3.4 Ga, ~20 % (upper 410 km) during ~3.6–2.7 Ga and 40–50% (upper 800–1,000 km) from ~1.8 Ga. The model is consistent with Re/Os, Rb/Cs, Nb/U and Th/U relations in mid-ocean ridge basalts (MORB) and ocean island basalts OIB.

5.5 The Re-Os System

The β-decay of ^{187}Re to ^{187}Os allows isotopic age determinations and petrogenetic estimates of the composition of precursor materials based on the relation (Shirey and Walker 1998):

$$^{187}Os/^{188}Os_t = {^{187}Os}/{^{188}Os_i} + {^{187}Re}/{^{188}Os_{chon}} \left(e^{\lambda(4.558 \times 10^9)} - e^{\lambda t} \right)$$

$$\left(^{187}Os/^{188}Osi = 0.09531 \right) \left(i - IIIA \ irons \right) \left(\lambda = 1.64 \times 10^{-11} \ year^{-1} \right)$$

$$^{187}Re/^{187}Os \ half-life \left(41.6 \times 10^9 \ year \right)$$

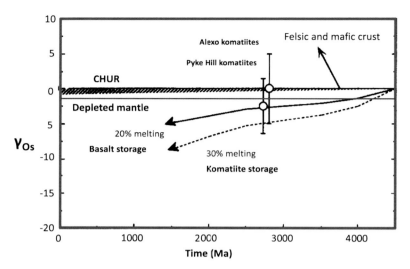

Fig. 5.5 Decrease in the mantle initial $^{187}Os/^{188}$ (γOs) with time as represented by komatiites and basalts, reflecting the consequence of preferential partitioning of Re to magmas and the crust (From McCulloch and Bennett 1994; Elsevier by permission)

Re-Os relations in iron meteorites indicate core crystallization in asteroids within the first 10–40 Ma of evolution of the Solar System. Overall the Earth mantle largely maintained a chondritic Re/Os pattern (Shirey and Walker 1998) including evidence of addition of siderophile elements during late accretion. Due to the small total volume of continental crust relative to the mantle and high Os levels of the mantle the fractionation of the continental crust resulted in little change in the $^{187}Os/^{188}Os$ of the mantle, with local exceptions (McCulloch and Bennett 1994) (Fig. 5.5). Mafic and felsic magmatic rocks display similar Re/Os ratio, indicating little intra-crustal fractionation of the refractory platinum group elements (Allegre and Luck 1980; McCulloch and Bennett 1994).

The differential compatibility of Re and Os relative to mafic phases results in elevated Re/Os ratios in mantle-derived magmas, rapid evolution of $^{187}Os/^{188}Os$ in mafic crustal rocks and low $^{187}Os/^{188}Os$ ratios in mantle residues (McCulloch and Bennett 1994). Consequently mantle samples such as komatiites are depleted in Re and have low $^{187}Os/^{188}Os$ (Fig. 5.5).

Chapter 6
Geochemical Trends of Archaean Magmatism

Abstract Comprehensive studies of well-exposed supracrustal sequences, including the Barberton Greenstone Belt, eastern Kaapvaal Craton, and Pilbara greenstone belts, northwestern Western Australia, allow detailed resolution of the relations between volcanic stratigraphy, isotopic ages and geochemistry, revealing systematic evolutionary trends of mantle and crust. This includes long term ~3.5 to ~3.2 Ga depletion of the source mantle in high field strength elements (HFS – Ti, Nb, rare earth elements) and a decrease in the ratio of Light REE to the heavy REE, as well as shorter term cycles displaying depletion and in some instances enrichment in the HFS elements, the latter likely implying juvenile mantle increments. Post 3.2 Ga basalts commonly display stronger LREE/HREE fractionation and Al-depleted compositions represented by high CaO/Al_2O_3 and TiO_2/Al_2O_3, implying increased importance of garnet in mantle residues and thereby cooler higher P/T (pressure/temperature) regimes. Felsic volcanic sequences ranging from andesites and dacites at low stratigraphic levels to rhyolites and K-rhyolites at higher levels represent increasingly fractionated compositions with time. Similar trends are shown by Archaean plutonic suites, an example being the increasing importance of garnet fractionation with time from ~3.55 to ~3.45 Ga implied by an increase in Sr and therefore the plagioclase component in TTG magmas (Moyen JF, Stevens G, Kisters AFM, Belcher RW, TTG plutons of the Barberton granitoid-greenstone terrain, South Africa. In: Van Kranendonk MJ, Smithies RH, Bennett VC (eds) Earth's oldest rocks. Developments in Precambrian geology 15. Elsevier, Amsterdam, pp 607–667, 2007).

Keywords Geochemical trend • Trace element • Source mantle • Rare earth elements • High field strength elements

6.1 Mafic and Ultramafic Volcanics

The unique association of tholeiitic basalts (TB), high-Mg basalts (HMB) (basaltic komatiites) and peridotitic komatiites (PK) in Archaean greenstone belts, representing high-degree melting of mantle sources as compared to post-Archaean periods, allows insights into the composition, evolution and petrogenetic processes of the Archaean mantle (Sun and Nesbitt 1977, 1978; Glikson 1983). The field of PK may be continuous with, or distinct from, the continuous fields of HMB and TB (Fig. 6.1). Where a compositional gap occurs it is possible that, whereas the HMB-TB suite formed by partial melting of the Archaean asthenosphere, PK magmas represent large-scale melting in ascending mantle plumes. Assuming steep Archaean geotherms of >25 °C/km and 20–40 % partial melting in a ~50 km-deep low velocity zone, followed by low pressure fractionation, a continuum of HMB – TB magmas may be produced. Episodic tectonic and/or impact events (Green 1972) could trigger ascent of mantle diapirs from levels of ~200 km and below, involving >50 % partial melting producing PK magmas. Low pressure fractionation of HMB – TB melts is supported by their commonly quartz-normative compositions (Fig. 6.1), suggesting breakdown of orthopyroxene under low pressure of <15 kb. Low-pressure melting and fractionation is supported by the scarcity to near-absence of alkaline rocks in most Archaean greenstone belts. Some similarities are noted between Si-enriched HMB and boninites found in ophiolites and fore-arc systems, interpreted in terms of hydrous mantle melting and/or involvement of continental lithosphere (Cameron et al. 1979; Smithies 2002).

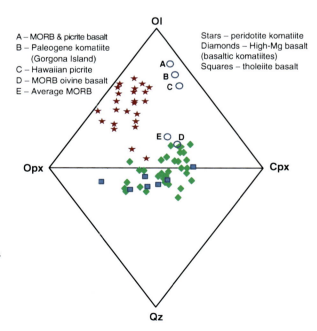

Fig. 6.1 Ol-Opx-Cpx-Qz ternary diagram displaying the fields of peridotitic komatiites (*stars*), high-Mg basalts (*diamonds*) and tholeiitic basalts (*squares*), compared to the compositions of (**a**) MORB picritic basalt; (**b**) Paleogene komatiite; (**c**) Hawaiian picrite; (**d**) Average MORB; (**e**) MORB olivine basalt

6.1 Mafic and Ultramafic Volcanics

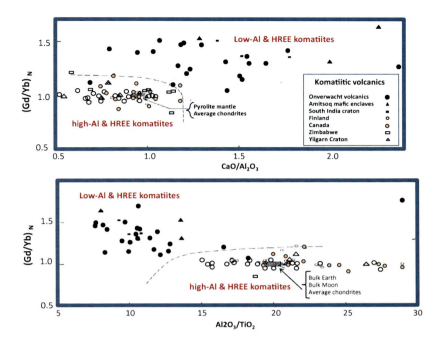

Fig. 6.2 Normalized rare earth element $(Gd/Yb)_N$ ratios vs CaO/Al_2O_3 and Al_2O_3/TiO_2 ratios in komatiites and basalts from early (pre-3.2 Ga) Archaean terrains (Onverwacht Group, Amitsoq gneiss mafic enclaves, Indian Craton) and upper to late Archaean terrains (Finland, Zimbabwe, Superior Province, Yilgarn Craton), indicating heavy REE-depleted and Al-depleted compositions of early Archaean mafic rocks, attributed to fractionation of residual garnet (From Jahn et al. 1981, 1982; Springer, by permission)

Sun (1987) developed a multistage accretion model of the Earth based on element abundance ratios in komatiites and ultramafic xenoliths. Mostly element distribution patterns are close to chondritic, with the exception of CaO/Al_2O_3 and V/Ti ratios, interpreted in terms of fractionation of aluminous garnet and of incorporation of vanadium in the core, respectively. A coexistence of Al-depleted and Al-undepleted komatiites attests to heterogeneous partial melting at different levels in a layered mantle, in agreement with a bimodal model of HMB-TB and PK derivation. A similarity between the ratios of siderophile and lithophile element (Ni/Mg, Co/Mg, Fe/Mg and P_2O_5/TiO_2) and PGE abundances estimated for the Archaean and modern mantle argue against continuous growth of the Earth's core.

Jahn et al. (1981, 1982) classified komatiites in terms of their rare Earth element (REE) patterns (Fig. 6.2):

Group I: HREE flat, $(Gd/Yb)N \sim 1.0$
Group II: HREE depleted, $(Gb/Yb)N > 1.0$; $CaO/Al2O3 > 1.0$; $Al2O3/TiO2 < 15$
Group III: HREE enriched, $(Gd/Yb)N < 1.0$; $CaO/Al2O3 < 1.0$; $Al2O3/TiO2 > 15$

The composition of Group II komatiites, prevalent in Early to mid-Archaean (~3.5–3.2 Ga) greenstone belts (Kaapvaal, India, southwest Greenland enclaves), signifies fractionation of garnet and thereby separation of Al_2O_3 and Yb from mafic and ultramafic partial melts. Fractionation of garnet would occur where the mantle solidus is intersected at depth ⩾200 km, hinting at high-pressure mantle source environments. By contrast, komatiites from Late Archaean (~3.0–2.6 Ga) greenstone belts (Finland, Canada, Yilgarn, Zimbabwe) display little HREE depletion and are more similar to chondrite composition. Thus, komatiites in the ~3.52–3.23 Ga East Pilbara terrain include both Al-depleted and Al-undepleted types, whereas those in the post-3.2 Ga West Pilbara terrain are typically Al-undepleted (Smithies et al. 2007a, b).

A comprehensive geochemical study conducted by Hallberg (1971) has shown volcanics of the Kalgoorlie-Norseman greenstone belts in the Yilgarn Craton are dominated by tholeiitic pillow basalts, displaying similarities to Archaean tholeiites from Canada and South Africa with respect to their low content of K, Rb, and Sr and Fe_2O_3/FeO ratios. However, unlike calc-alkaline volcanic cycles of the Superior province, andesites are rare and only limited geochemical cyclicity is manifest in Yilgarn volcanic sequences.

The well–exposed low metamorphic grade greenstone belts of the Barberton Greenstone Belt eastern Kaapvaal Craton (Glikson and Jahn 1985) and the Pilbara Craton of northwestern Australia (Glikson and Hickman 1981; Glikson 2007, 2008; Smithies et al. 2005, 2007a, b) allow detailed resolution of temporal geochemical trends of mafic-ultramafic volcanic sequences. Pre-3.2 Ga greenstone sequences of the eastern Pilbara Craton display a decline in TiO_2, REE and light/heavy REE ratios during ~3.52–3.23 Ga (Figs. 6.3 to 6.5). By contrast post-3.2 Ga mid-late Archaean basalts display a wide compositional range including high La/Sm and Sm/Yb ratios, signifying sharp increase in LREE/HREE fractionation associated with development of semi-continental environments (Glikson and Hickman 1981; Smithies et al. 2005, 2007a, b) (Figs. 6.3 and 6.4), likely implying variations in the role of garnet fractionation and thereby of pressures in mantle source regions. Smithies et al. (2007a, b) identified intercalated low-TiO_2 and high-TiO_2 basalts (TiO2 >0.8 %), the latter showing high levels of high field strength elements (Hf, Zr, Ti, Nb, Ta), high rare earth elements (REE) levels and Gd/Yb ratios (1.12–2.23), high Fe, low Al_2O_3/TiO_2 ratios (18.7–8.9).

Superposed on the long-term depletion trends in HFSE are smaller scale cycles. Basalt sequences of the North Star Basalt (NSB) and Mount Ada Basalt (MAB) (Fig. 6.5) display systematic depletion in TiO_2 with stratigraphic levels, suggesting depletion of the source mantle in HFSE elements (Glikson and Hickman 1981). A concentration of volatiles, alkali elements, S and Cu toward the top of MAB sequence is attributable to late syn-volcanic carbonization or/and secondary upward migration of volatiles associated with leaching of mafic volcanic rocks. The oldest volcanics of the North Star Basalt (NSB ~3,490 Ma) consist of Qz-normative HMB closely interspersed with tholeiitic basalts and dolerites with high Ti and low K, Al, Ni and Cr levels. High Ti and low Ni and Cr basalts of the NSB are distinct from high-Ni and Cr modern mid-ocean ridge basalts. The overlying MAB is locally dominated by heavily carbonated tholeiitic basalts and dolerites. In both the eastern and western Pilbara the mantle sources of the basalts were typically less depleted in

6.1 Mafic and Ultramafic Volcanics

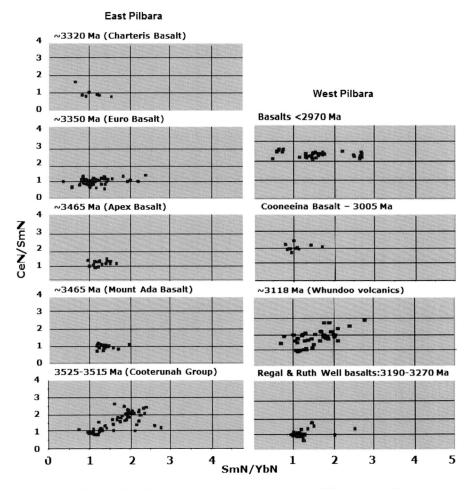

Fig. 6.3 Ce$_N$/Sm$_N$ vs Sm$_N$/Yb$_N$ plots for basaltic volcanic units, Pilbara Craton (Data sources: Glikson and Hickman 1981; Glikson 2007, 2008; Smithies et al. 2005, 2007a, b)

incompatible elements than normal MORB sources (Smithies et al. 2007a, b). By contrast to these ~3.49 Ga basalts, mafic and ultramafic volcanics of the ~3.35 Ga Euro Basalt display enrichment in TiO$_2$ with stratigraphic level (Fig. 6.6), likely reflecting the tapping progressively enriched juvenile mantle sources, possibly mantle plumes.

Trace element characteristics of TB of different stratigraphic units commonly correlate with those of associated high-Mg basalts and peridotitic komatiites, suggesting primary magmatic inheritance of these features. Two geochemical series can be discerned based on Ni and Ni/Mg: (a) High-Ni low-Mg/Ni basalts associated with komatiites; (b) low-Ni high-Mg/Ni basalts of commonly low Mg' values (Mg/Mg + Fe) accompanied by minor komatiite component. Trace metal indices (V/Cr,

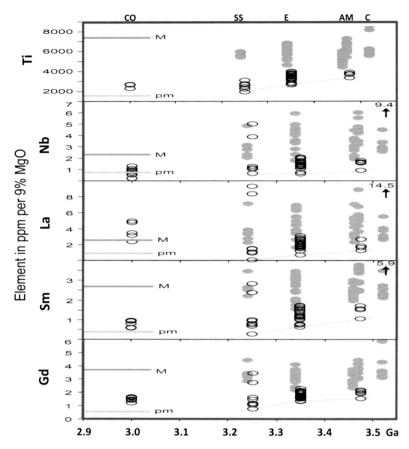

Fig. 6.4 Variations in the compositions of high-Ti basalts (*grey circles*) and low-Ti basalts (*open circles*) of the Pilbara Supergroup with time. *C* Coonterunah Subgroup, *M* Mount Ada Basalt, *A* Apex Basalt, *E* Euro Basalt, *SS* Sulphur Springs Group, *CO* Coonieena Basalt Member (Bookingarra Formation, Croydon Group, De Grey Supergroup). Values for N-MORB (*M*) and primitive mantle (*pm*) are from Sun and McDonough (1989). Trace element concentrations recalculated to values expected at 9 wt.% MgO (From Smithies et al. 2007a; GSWA, by permission)

Co/Ni, Ti/Ni) suggest high fractionation of ~3,490 Ma TB, lower fractionation of ~3,449 Ma TB; low-to-intermediate degree of fractionation of post-3,225 Ma TB; variable fractionation of 2.78–2.63 Ga Fortescue Group plateau basalts (Glikson et al. 1989). Petrogenetic model calculations suggest somewhat siderophile elements-enriched mantle source compositions. As suggested by compositional gaps between PK and HMB, these magmas may not be generally inter-related by crystal fractionation. HMB likely formed by 30–50 % melting of mantle peridotite in shallow Archaean equivalent of the modern low velocity zone. The continuous chemical spectrum between high-Mg basalts and tholeiitic basalts suggests the latter could form from primary high-Mg magma by 25–65 % crystal fractionation,

6.1 Mafic and Ultramafic Volcanics

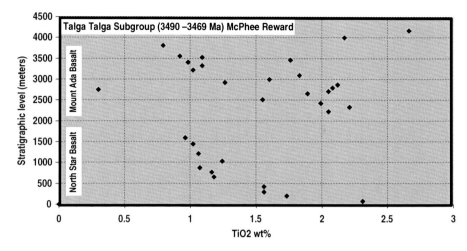

Fig. 6.5 Variations in TiO$_2$ in basalts with stratigraphic levels: North Star Basalt (~3.49 Ga) and Mount Ada Basalt (~3.47 Ga), Marble Bar region, Pilbara Craton

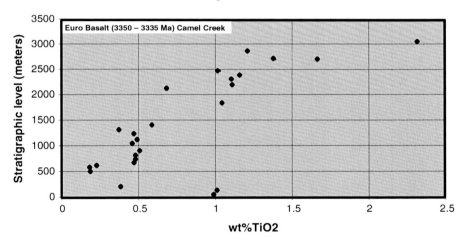

Fig. 6.6 Variations in TiO$_2$ in basalts with stratigraphic levels: Euro Basalt (~3.35–3.325 Ga) Camel Creek, Marble Bar region, Pilbara Craton

leaving residues of olivine, clinopyroxene, orthopyroxene and plagioclase. The high-K values of post-3,225 Ma tholeiitic basalts may represent effects of contamination by felsic crust.

Some of the geochemical trends indicated earlier in this section suggest a secular depletion of the mantle in incompatible elements in some regions of the Archaean Earth, consistent with observations by Griffin et al. (2003) who indicated a systematic variation in the composition of the subcontinental lithospheric mantle with age. These authors indicate the Archaean subcontinental lithospheric mantle is distinctively different from younger mantle and is highly depleted in incompatible elements, containing depleted sub-calcic harzburgites which are essentially absent in younger sub-continental mantle. On the basis of this trend and isotopic and trace element geochemistry of mantle xenoliths these authors suggest a quasi-contemporaneous formation or modification of the crust and its underlying mantle root zones, implying the crust and mantle have in many cases been coupled in time and space through their subsequent history. A further development of the theory conveyed by Griffin et al. (2014) based on worldwide compilation of U/Pb, Hf-isotope and trace-element data on zircon, and Re–Os model ages on sulfides and alloys in mantle-derived rocks and xenocrysts develops a 4.5–2.4 Ga mantle and crust model invoking long periods of stagnant mafic crustal conditions interrupted by pulses of juvenile magmatic activity at ca 4.2 Ga, 3.8 Ga and 3.3–3.4 Ga the authors interpret in terms of mantle plume episodes on a time scale of ~150 Ma. In this model generation of continental crust peaked in two main pulses at ~2.75 Ga and ~2.5 Ga. However, as indicated in Chaps. 2, 4, 7, 8 and 9 extensive magmatic activity took place in several cratons during intervening periods.

6.2 Felsic Igneous Rocks

Geochemical studies of Archaean granitoids (Arth and Hanson 1975; Hanson 1975; Arth 1976; Arth et al. 1978) display clear trends from Na-rich syntectonic magmas to fractionated K-rich magmas (Glikson and Sheraton 1972; Glikson 1976a, b, c, 1979a; Glikson and Hickman 1981; Martin 1999; Smithies 2000; Smithies and Champion 2000; Smithies et al. 2003; Champion and Smithies 1999, 2007; Martin et al. 2005; Moyen et al. 2007, 2009) (Figs. 6.7 to 6.11), underpinning the significance of a temporally distinct Archaean Tonalite-Trondhjemite-Granodiorite (TTG) suite (Barker 1979). Analogous dacitic compositions pertain to Archaean felsic volcanic and hypabyssal porphyries (O'Beirne 1968). Archaean calc-alkaline volcanic sequences commonly display evolution from dacites to K-rhyolites, as recorded in the ~3.47 Ga volcanics of the Duffer Formation, Pilbara Craton (Glikson and Hickman 1981) (Fig. 6.7).

Archaean batholiths include multiple generations of intrusive bodies forming a complex geometry where in some instances younger plutons penetrate at the centers of older plutons and in other instances post-kinematic intrusions are emplaced at the margin along contacts with greenstone belts (Fig. 6.8). Geochemical data for granitoids from Archaean granite-greenstone terrains and gneiss-granulite terrains, classified on the An-Ab-Or ternary (O'Connor 1965) display continuous compositional spectrum between the fields of tonalite, trondhjemite, granodiorite and adamellite. Tonalites dominate among pre-3.2 Ga granitoids of the Kaapvaal Craton (Glikson 1976a, b, 1979; Moyen et al. 2007; Moyen and Martin 2012) (Figs. 6.8 and 6.9). By contrast ~3.3 Ga Pilbara granitoids display a wide range of K_2O/Na_2O ratios, including relatively differentiated compositions (Moyen et al. 2007). The tonalitic chemistry of Barberton plutons, as distinct from Western Australian trondhjemites and

6.2 Felsic Igneous Rocks

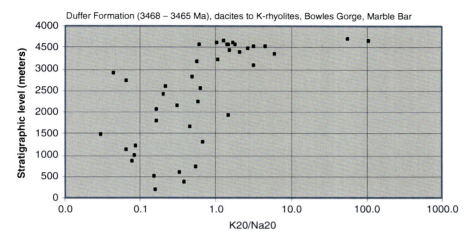

Fig. 6.7 Stratigraphic variations of the K/Na ratio in a dacite to K-rhyolite sequence of volcanic rocks and pyroclastics of the Duffer Formation (~3.468–3.465 Ga), Bowles Gorge, Marble Bar area, Pilbara Craton

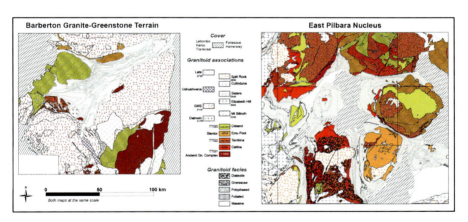

Fig. 6.8 Granitoids in the East Pilbara craton and the Barberton Granite – Greenstone terrain record different modes of Archaean crustal accretion (From Moyen et al. 2011; courtesy J-F Moyen)

granodiorites, may be attributed to an ultramafic composition of the precursor materials, consistent with the komatiite-dominated composition of the Onverwacht Group.

Felsic igneous rocks associated with the Barberton Greenstone Belt, Kaapvaal Craton (Glikson 1976a, b), display low abundances of large ion lithophile (LIL) elements pertaining to the 'ancient tonalites' plutonic suite which intruded the volcanics at ~3.4–3.2 Ga and to albite porphyries intercalated with early ultramafic-mafic volcanics of the Onverwacht Group. 3,548–3,298 Ma felsic volcanic units display REE patterns similar to those of the intrusive tonalites, with total REE, Ce_N/Yb_N and

Rb/Sr values comparable to some tonalitic-trondhjemitic components of the Ancient Gneiss Complex in Swaziland (~3,644–3,433 Ma, Compston and Kröner 1988). Some of the felsic volcanics display highly fractionated REE patterns suggesting equilibration with garnet and thereby origin by high-pressure melting of eclogite at depths of 30–50 km. REE patterns of the 'ancient tonalites' are characterized by pronounced positive Eu anomalies, reflecting the abundance of plagioclase. Some of the albite porphyries may represent shallow-level hypabyssal equivalents of the 'ancient tonalites'. Late-stage members of the 'ancient tonalites' are present, examples being the Dalmein and Bosmanskop plutons, representing decreasing partial melting of basic source rocks. In the Barberton terrain post tectonic K-rich adamellite and migmatite are represented by the 3.0 Ga Hood granite and Nelspruit migmatite, displaying field evidence for derivation by anatexis of the 'ancient tonalites' as well as incorporation of ultramafic to mafic xenoliths derived from the Onverwacht Group.

Orthogneisses from high-grade Archaean terrains tend to be more mafic than samples from granite-greenstone terrains, likely due to extensive digestion of mafic materials. Apart from high FeO/MgO gneisses such as the Amitsoq gneisses (Lambert and Holland 1976), the Fe-Mg-Alk relations of Archaean Na-rich granitoids define calc-alkaline trends, similar to post-Archaean Na-rich granitoids, with the notable exception of oceanic plagioclase-rich granites (Coleman and Peterman 1975). The differentiation Indices (Qz + Or + Ab + Ne + Le + Kp) and Solidification Indices (100 MgO)/ (MgO + FeOt + Na$_2$O + K$_2$O) of Archaean gneisses confirm the

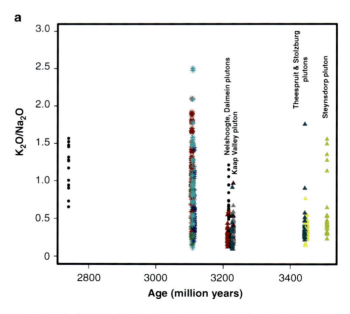

Fig. 6.9 (a) Variations in K$_2$O/Na$_2$O with time in granitoids of the Barberton Mountain Land, eastern Kaapvaal Craton. Courtesy J-F Moyen. (b) K$_2$O/Na$_2$O vs. Sr diagrams for Barberton and Pilbara rocks. Barberton samples on the *left*, Pilbara on the *right*. Samples are grouped by chronological order, *bottom* to *top* (From Moyen et al. 2009; Royal Society of Edinburgh, by permission)

6.2 Felsic Igneous Rocks

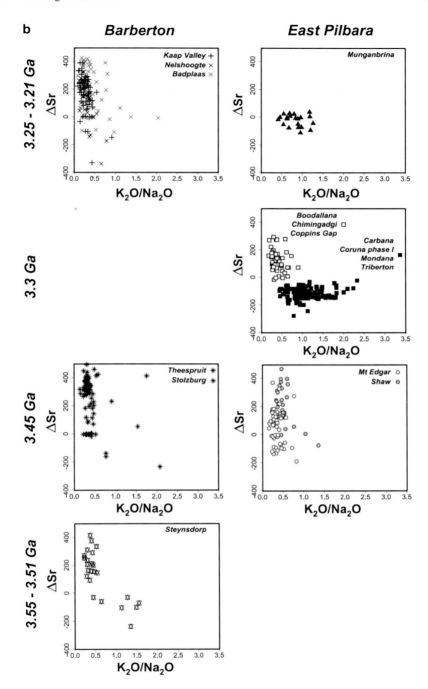

Fig. 6.9 (continued)

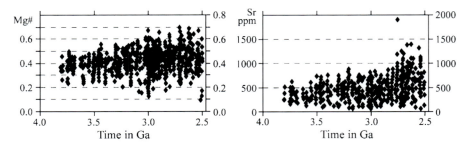

Fig. 6.10 Diagrams showing time-evolution of Mg# and Sr in TTG magmas during the Archaean. The upper envelope of the group of points represents the composition of the more primitive TTG magmas (From Martin and Moyen 2002; Elsevier by permission)

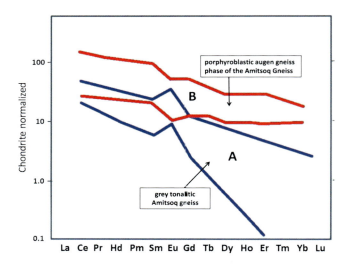

Fig. 6.11 Contrast between the REE patterns of tonalitic gneisses and K-granites, an example from southwest Greenland. (**a**) Grey (tonalitic) Amitsoq gneiss; (**b**) porphyroclastic augen gneiss phase of the Amitsoq Gneiss (From O'Nions and Pankhurst 1978; Elsevier, by permission)

relatively basic composition of Barberton and Minnesota – Ontario samples, whose FeO/MgO ratios are lower than for Western Australian rocks. Some Yilgarn trondhjemites are of siliceous, high-alkali and ferromagnesian-poor chemistry, exemplified by granitoid pebbles in Kurrawang conglomerate. Relatively ferroan compositions are associated with low Na_2O/K_2O and low CaO. The alumina levels of Archaean granitoids are low to moderate (13–16 % Al_2O_3) and with sediment-derived paragneisses are rare.

Trace element abundances in Archaean granitoids show marked fluctuations of the large ion lithophile (LIL) elements, including Sr, Ba and Rb. Some Yilgarn, Kaapvaal and Minnesota and Ontario TTG have remarkably high Sr (<800 ppm) contrasted with Pilbara trondhjemites with Sr abundances lower by factors of 2–4. Positive Sr-CaO correlations (r = 0.70, 0.61) pertain to Pilbara and Minnesota-Ontario compositions. Many samples from high-grade Archaean terrains have high Ba abundances as compared to data from granite – greenstone terrains. Ba decreases from ~3.7 to 3.6 Ga granitoids to ~3.0–2.8 Ga granitoids. High Ba and Sr levels pertain to some Minnesota-Ontario syeno-diorites and Barberton plutons of alkaline affinities (Arth and Hanson 1975; Glikson 1976a, b). Rb levels display a wide dispersion, as shown for example by comparisons between Minnesota-Ontario tonalites (Rb ~ 34 ppm), Barberton tonalites (Rb ~ 42 ppm), Pilbara trondhjemites (Rb ~ 93 ppm) and Kurrawang trondhjemite pebbles (Rb ~ 12 ppm). Gneisses of the Swaziland terrain have higher Rb levels than Barberton tonalites, a difference to which Hunter (1974) referred as evidence against their coeval relations. Lewisian gneisses may be extremely depleted in Rb, with the average of 254 Drumbeg gneisses with 11 ppm Rb (Sheraton 1970). Higher levels pertain to Nuk gneisses and the older Amitsoq gneisses have an average of 71 ppm for 25 samples. With few exceptions, such as the Lewisian gneisses, Archaean felsic plutonic rocks have low to moderate K/Rb ratios of below 400. Trondhjemite and granite tend to have higher Rb/Sr ratios (0.5–1.0) than tonalites (<0.15).

Hurst (1978) studied the evolution of initial $^{87}Sr/^{86}Sr$ (Ri) ratios between 3.9 and 1.6 Ga, deducing an average mantle Rb/Sr ratio of 0.014, implying early depletion of the mantle in Rb, likely selectively transferred to mafic and felsic crust, a conclusion consistent with the decrease in Rb/Sr from ~3.7 to 3.6 G Amitsoq and Uivak gneisses to ~3.0–2.8 Ga Nuk and Lewisian gneisses. A temporal increase in the Rb abundance of Archaean granites was reported from the Barberton – Swaziland terrain (Viljoen and Viljoen 1969a, b; Condie and Hunter 1976) and from the Yilgarn Craton (Glikson 1979). The low initial Sr^{87}/Sr^{86} (R_i) values in Archaean magmatic bodies (Hurley et al. 1962; Moorbath et al. 1976; Moorbath 1977) (Fig. 5.2) constrain the crustal prehistory and suggest derivation from juvenile mantle-derived materials. Other views suggest continuous recycling of SIAL crust associated with secondary depression of Ri values through isotopic exchange with the mantle ('zone refining') (Armstrong 1968). The significance of Ri values for crustal residence times of precursor materials depends on assumptions regarding the nature of these precursors and their Rb/Sr ratios as well as assumptions regarding the mantle Ri growth curve. Secondary depression of the Ri values depression has also been suggested as a result of crustal recycling (Bridgwater and Collerson 1976, 1977) and volatile activity associated with mantle degassing (Collerson and Fryer 1978), a suggestion requiring special pleading. Thus, recycling of granitoid crust would result in isotopic age resetting, contrasted with the preservation of an age record as far back as ~3.8 Ga (Black et al. 1971; Moorbath et al. 1973). The low Ri values of tonalites and trondhjemites in high-level TTG bodies subject to little or no 'zone refining' places strict limits on crust-mantle recycling processes.

Compilations of Rb-Sr ages and Ri values indicate low fractionation of Archaean TTG (Fig. 5.2). In the Zimbabwe Craton plutonic rocks include (1) ~3.6–3.4 Ga tonalitic gneisses of low Ri values (0.700–0.702); (2) ~3.0 Ga tonalite – trondhjemite – granodiorite plutons of low Ri values; (3) ~2.8–2.7 Ga plutons of low Ri values; (4) ~2.7–2.6 Ga intrusions including (a) small bodies of tonalite to granite of low Ri values which intrude Bulawayan Group greenstones (Somabula and Sesombi tonalites) and (b) extensive high-level adamellite – granite yielding intermediate Ri values (0.704) (Chilimanzi and Zimbabwe batholiths); (5) high values such as 0.711 occur in the Mont d'Or gneiss (Pb-Pb age – 3,345 ± 55 Ma) (Moorbath et al. 1976; Taylor et al. 1991). Intermediate values of ~0.705–0.706 are recorded in Swaziland gneisses where a linear correlation occurs between Ri values and age (Davies and Allsopp 1976). It appears that, where two-stage mantle melting processes (Green and Ringwood 1967) dominated formation of felsic magmas, further intra-crustal melting took place at younger stages of crust formation. In the Pilbara Craton Rb-Sr isochron ages reported for the Mount Edgar and Shaw batholiths comprise 3.1–2.9 Ga (Ri ~0.702) ages of older gneiss phases and ~2.7–2.6 Ga (Ri ~0.735) ages for post-tectonic granites (DeLaeter and Blockley 1972; DeLaeter et al. 1975). U-Pb zircon ages of gneisses from the Mount Edgar and Shaw batholiths yielded ages of about 3.4–3.3 Ga (Pidgeon 1978) whereas Pb-Pb K-feldspar ages of Pilbara gneisses appear to be reset at 3.0–2.9 Ga (Oversby 1976).

The high field strength lithophile (HFSL) elements (Th, Pb, Zr, Nb) are depleted in some Archaean gneisses, for example in Barberton tonalites and in southwest Greenland (Lambert and Holland (1976). Two distinct groupings apply to Yilgarn granites, one represented by trondhjemites (Kambalda pluton: $U = 1-3$ ppm; $K_2O = 2.1-2.4$ %) and the other by high-K granites with up to 10 ppm U. Some Pilbara samples have anomalously high U values of about 15 ppm, possibly reflecting secondary enrichment. Good correlations are observed between Pb and K_2O. (Yilgarn $r = 0.67$; Pilbara $r = 0.82$). Some Amitsoq gneisses have high Zr levels (200–400 ppm) in comparison to other gneisses of the North Atlantic Craton, with a main range of 100–250 ppm Zr. Guiana gneisses define two Zr-TiO_2 groupings, one with 130–230 ppm Zr and 0.25–0.40 % TiO_2 and the other with 20–150 ppm Zr and 0.20–0.70 % TiO_2. Nb values of Archaean rocks appear to be low compared to the average high-Ca granite of Turekian and Wedepohl (1961) (Nb = 20 ppm). A good correlation of Nb with Zr is apparent for the Amitsoq gneiss ($r = 0.76$).

Quartz diorites, tonalites and trondhjemites ~2.7 Ga-old from northeast Minnesota and northwestern Ontario have steep chondrite-normalized rare earth (REE) patterns and, in some instances, positive Eu anomalies (Arth and Hanson 1975). Some samples display extreme depletion in heavy REE, with Yb_N levels of <1.0 and in some instances <0.1, attributable to separation of mineral phases with high partition coefficients ($^{Gnt}K_{Yb} = 39.9$ for dacitic melts) and/or hornblende ($^{Hbl}K_{Yb} = 4.89$ for dacitic melts) (Arth 1976). By contrast positive Eu_N anomalies in tonalitic magmas represent low partition coefficients of Eu in garnet and hornblende.

By contrast to the TTG suite REE patterns of post-tectonic quartz monzonite show negative Eu anomalies, attributed to plagioclase fractionation, and in many instances show low degrees of fractionation of HREE. The K-rich intrusions display eutectic low-melting compositions, high LIL element levels and highly variable Ri values (Arth and Hanson 1975; Condie and Hunter 1976; Glikson 1976a, b; Hunter et al. 1978). The plutons occur either as discrete late high-level intrusions in granite – greenstone terrains or as zones and bands of porphyritic gneiss in high-grade terrains. The geochemical characteristics of the K-rich bodies have been interpreted in terms of partial melting of greywackes (Arth and Hanson 1975), siliceous granulite (Condie and Hunter 1976) or of older TTG basement (Glikson 1976a, b), consistent with field relations in South Africa and Western Australia. A third plutonic class identified by Arth and Hanson (1975) comprises syeno-diorites and syenites characterized by high total REE, strong LHEE enrichment ($Ce_N = 200-500$), fractionated REE $(CeN/Yb)_N = 30-60$) and a lack of Eu anomalies. Similar characteristics are displayed by the Bosmanskop pluton, a late-stage member of the Barberton suite intruded into the Ancient Tonalites (Anhaeusser 1974; Glikson 1976a, b).

The trace element and isotopic parameters outlined above suggest derivation of TTG by partial melting of eclogite, basic granulite or amphibolite (O'Nions and Pankhurst 1974; Condie and Hunter 1976; Glikson 1976a, b; Hunter et al. 1978; Collerson and Fryer 1978; Drury 1978; Collerson and Bridgwater 1979). Arth and Barker (1976) emphasized the role of amphibole fractionation in preference to garnet. REE patterns allow estimates of degrees of partial melting, for example for the Saganage Tonalite (Ce/Yb ~ = 12–54; Eu/Eu* = 0.87–1.90), modelled as the product of 30–35 % melting of quartz eclogite (30 % gnt, 55 % cpx, 15 % qz) (Arth and Hanson 1975). An alternative interpretation regards the HREE depletion of TTG as a result of complexing and leaching of REE elements by CO_2, SO_2, Cl and F bearing fluids (Collerson and Fryer 1978). However, this model does not account for the systematic field and petrological differences preserved between the TTG suite and late K-rich granitoids as indicated above. Lambert and Holland (1976) pointed out the generally low yttrium abundances (Y ~ 10–30 ppm) and low Y/CaO ratios in Archaean volcanics as compared with modern ocean-floor tholeiites (Y ~ 20–60 ppm). Low Y/CaO ratios are well pronounced in Lewisian gneisses (Sheraton 1970; Glikson and Sheraton 1972). The low HREE and Y may be attributed to depletion of the Archaean mantle in HREE and Y or, alternatively, arrest of HREE and Y in residual amphibole or garnet.

The field relations, geochemical and isotopic data reviewed above provides boundary conditions for models of origin of Archaean batholiths. Tonalitic and trondhjemitic magmas are unlikely to be the product of anatexis of granitoid basement rocks as the typical products of anatexis of felsic rocks yield adamellite to granite magmas (Carmichael et al. 1974). An exception would be where total melting or remobilization of TTG crust occurs, requiring superheating and extreme H_2O pressures, such as have yet to be demonstrated. Field records in high grade metamorphic terrains indicate re-melting of TTG granitoids results in segregation of K-rich leucosome fractions from paleosome restite. Thus models advocating formation of the TTG suite by anatexis of older granitic crust (Baragar and McGlynn 1976; Archibald et al. 1978) do not account for the geochemical differences between the TTG suite and the K-rich LIL element-high products of en-SIAL anatexis.

Viljoen and Viljoen (1969c) regarded the Barberton tonalites as products of mantle melting and Taylor (1967) interpreted the origin of Archaean batholiths in terms of large-scale up-streaming of lithophile elements from the mantle. Low-degree melting of hydrated peridotite under pressures of ~10 kbar is capable of yielding basaltic andesite (Kushiro 1972; Mysen and Boettcher 1975). However, whereas small volumes of dacite may ensue by fractional crystallization of mantle-derived andesitic melts, it is unlikely the large volumes of the TTG suite could arise in this manner. In this regard it is pertinent to note the relative paucity of andesites in older Archaean greenstone belts, as contrasted with their appearance in late greenstone belts in the Superior Province (Folinsbee et al. 1968; Baragar and Goodwin 1969; Green and Baadsgaard 1971) and the Yilgarn Craton (Hallberg 1971).

Assuming a high geothermal gradient of ~40 °C/km in Archaean mafic-ultramafic crustal domains (Glikson and Lambert 1976), partial melting of H_2O-saturated olivine tholeiite should commence about 20 km depth and segregation of dacitic melts would be widespread between 20 and 30 km depth. Under anhydrous conditions this process would commence at ~35 km depth. Garnet becomes stable at subsolidus temperatures between 40 and 50 km, depending on partial H_2O pressure. Under pressure of about ~30 kbar the breakdown of amphibole results in a release of water and depression of the solidus, defining the low-velocity zone. However, under high geothermal gradients the breakdown of amphibole would take place under lower pressure, resulting in a shallower upper limit of the low-velocity zone. The bulk of partial melting would be concomitant with dehydration under high pressures associated with transformation of amphibolite to granulite and eclogite. Under very high geothermal gradients granulite facies conditions are attained before the H_2O-saturated solidus is intersected. In this case melting would only commence at or above 1,200 °C at depths of ~10–20 km. Addition of water to the magma during its ascent would, lowering the solidus, result in lower viscosity and enhance the assimilation of invaded supracrustals.

Apart from the melting of mafic crust it is possible re-melting of trapped basic pods within the Archaean upper mantle and fractional crystallization of basic magmas have contributed to formation of the TTG suite. On the other hand, fractional crystallization would result in an abundance of rocks of intermediate composition, such as are comparatively rare in early Archaean greenstone belts and in Archaean batholiths. Questions regarding two-stage mantle melting model include: (a) whereas partly digested relics of assimilated mafic material are common in xenolith chains (Chap. 3), no examples are evident of ultramafic residues of the partial melting process at or below deep seated roots zones of TTG batholiths, for example in Lewisian granulites formed under 10–15 kbar (O'Hara 1977). Exceptions may be represented by spinel dunite and spinel pyroxenite enclaves in the Fiskenaesset region, southwest Greenland (Friend and Hughs 1977). Assuming that the volumes of residual refractory materials were at least equal to that of the TTG batholiths, it must be inferred the dense residues must have sunk into the Archaean mantle.

The nature of the structural environment and tectonic mechanism which facilitated partial melting of vast volumes of mafic-ultramafic crust remains little understood. Geochemical and isotopic parameters diagnostic of the composition of source materials and the nature of fractionation processes are not definitive with respect to identification of tectonic environments, except where uniformitarian assumptions are made (Chap. 12). Attempts at correlating trace element features of ancient volcanics with tectonic environments (cf. Floyd and Winchester 1975; Pearce et al. 1977) are complicated by likely lateral vertical and lateral mantle heterogeneities and temporal evolution of mantle composition (Sun and Nesbitt 1977; Jahn et al. 1979; Glikson and Hickman 1981), which cast doubt on the validity of geochemical-tectonic correlations. Contrasting schools of thought have developed during the last decades with regard to the interpretation of Archaean tectonic environments, including

(a) models advocating an early granitoid basement below greenstone belts (cf. Windley and Bridgwater 1971);
(b) plate tectonic-based models interpreting high-grade Archaean terrains in terms of assembly of coexistence of micro-cratons associated with subduction of mafic-ultramafic crustal tracts (cf. Bridgwater et al. 1974; Nutman et al. 2007);
(c) Non-uniformitarian crustal transformation models.

Models type A make a distinction between gneiss – granulite terrains and greenstone belts, the latter believed to form above the former. Models type B advocate horizontal tectonics, plate tectonics, subduction and lateral inter-thrusting of granitic and mafic crustal analogous of modern circum-Pacific regimes, notably the Chilean ophiolite – granodiorite complexes (Windley 1973; Bridgwater et al. 1974; Tarney et al. 1976; Windley and Smith 1976; Tarney and Windley 1977; Moorbath 1977). Models type C emphasize the antiquity of early ultramafic – mafic volcanic enclaves in the isotopically oldest-documented tonalite – trondhjemite batholiths, the absence of evidence for granitic basement directly underlying early pre-3.2 Ga mafic-ultramafic sequences (Onverwacht Group, Warrawoona Group), coeval transitions between granite – greenstone and deep-seated gneiss – granulite root zones, and vertical tectonics associated with diapiric plutonic emplacement of the gregarious batholiths (Glikson 1970, 1972, 1976a, b; Anhaeusser et al. 1969; Viljoen and Viljoen 1969c; Anhaeusser 1973; Glikson and Lambert 1976; Naqvi 1976; Weber and Scoates 1978).

Geochemical analogies between post-3.2 Ga greenstone belts of the Superior Province and low-K tholeiite-andesite-dacite-rhyolite assemblages of modern island-arcs (Folinsbee et al. 1968; Baragar and Goodwin 1969; Green and Baadsgaard 1971) support type-B models. Likewise, significant similarities are observed between Archaean and post-Archaean granite-greenstone terrains, such as the middle Proterozoic Snow Lake – Flin Flon terrain (Stauffer et al. 1975) and Phanerozoic ophiolite – granodiorite complexes (Tarney et al. 1976; Hietanen 1975a, b; Burke et al. 1976). However, these similarities, interpreted in terms of two-stage mantle melting processes (Green and Ringwood 1967, 1977), do not necessarily imply identical structural environments. Fundamental distinctions between Archaean terrains and younger orogenic belts include (Chap. 12):

1. Abundance of temporally unique peridotitic komatiites and high-Mg basalts in Archaean greenstone belts.
2. Dominance of the bimodal low-K tholeiite – Na dacite suite and paucity of andesite in pre-3.2 Ga Archaean greenstone belts.

3. The near-absence of alkaline volcanic and plutonic rocks within Archaean batholiths, except at final stages of their evolution.
4. Evidence for near-liquidus temperatures of the Archaean tonalite – trondhjemite granodiorite magmatic suite and the digestion of vast volumes of ultramafic – mafic crust by the felsic magmas.

The Na-rich felsic igneous rocks can be modelled in terms of equilibrium melting of mafic-ultramafic rocks of average high-Mg basalt composition and fractional crystallization of a mafic parent magma involving separation of garnet (Glikson and Jahn 1985). By contrast near-flat heavy REE profiles of post-tectonic K-rich granites (Lochiel Granite) display field evidence for derivation by anatexis of tonalite-greenstone crust about ~3.0–2.6 Ga, confirmed by distinct REE patterns including HREE depletion and Eu enrichment of tonalites as compared to K-rich granites (Fig. 6.11). Likewise siliceous gneisses of the Ancient Gneiss Complex of Swaziland with high total REE, high Ce_N/Sm_N, low Sm_N/Yb_N, negative Eu anomalies and high Rb/Sr can be modelled in terms of late or post-magmatic anatexis of the tonalites.

The distribution patterns of rare earth elements (REE), alkali, alkaline earth elements (Rb, Ba, Sr) and high field strength elements (Ti, Zr, Nb) in felsic volcanic and plutonic rocks of the Pilbara Block, Western Australia (Glikson et al. 1987) define a number of associations:

(A) 3.5–3.4 Ga Na-rich calc-alkali and occasionally high-Al andesite-dacite-rhyolite volcanics within greenstone belts and plutonic equivalents. This association is dominated by low Rb/Sr ratios (<0.4) moderate to high light/heavy REE fractionation, $(Ce/Yb)_N$<15, small or no Eu/Eu* anomalies and near-flat heavy REE profiles, $(Dy/Lu)_N = 0.9$–1.2.
(B) 3.32–2.9 Ga includes K-rich rhyolite lavas at high stratigraphic levels of the Warrawoona Group (Wyman Formation) and late to post-tectonic granites. This association includes rocks of very high Rb/Sr ratios (11–15), moderate $(Ce/Yb)_N$ (5–6), marked Eu/Eu* anomalies (<0.4) and flat heavy REE profiles ($(Dy/Lu)_N = 0.9$–1.1).
(C) ~2.9 Ga rhyolitic pyroclastics (Mons Cupri Volcanics) and related plugs (Mount Brown Rhyolite) of the Whim Creek Group. This association includes rocks of moderate Rb/Sr ratios (0.6–1.0), moderate $(Ce/Yb)_N$ (5.7–7.0), low Eu anomalies and weakly fractionated heavy REE $(Tb/Yb)_N$<1.0.

A derivation of magmas of association (A) by partial melting or fractional crystallization of basic rocks can be modelled by separation of residual clinopyroxene, plagioclase and minor garnet. This is consistent with the very low Nb anomalies and lack of Eu and Sr depletion in the andesite-dacite magmas, which suggests a minor role of plagioclase separation and thus unlikely derivation by anatexis of felsic crust. By contrast the origin of associations (B) and (C) can be modelled in terms of anataxis of tonalite/trondhjemite crust, possibly including a mafic component, involving separation of a residue of plagioclase and clinopyroxene. These models are consistent with an evolutionary trend from largely mafic to largely felsic sources with time. The low-LIL dacites of the Duffer Formation probably formed by partial melting of basic source materials, but late-magmatic hydrothermal addition of Si and K is required for the rhyolites. Elvan-type quartz porphyries of the Wyman Formation were related to late-stage hydrous differentiation. Smithies et al. (2007a, b) observe a trend toward silica and alumina enrichment and an increase in La_N/Yb_N in felsic volcanics with time in Pilbara greenstone belts. These authors point to compositional differences between felsic volcanics and the tonalite-trondhjemite suite and emphasise fractional crystallization of tholeiitic magma accompanied with crustal contamination, as well as an increase in the role of high-pressure melting with time.

Several petrological and geochemical studies of volcanic sequences of the Superior Province and Slave Province greenstone belts report (1) a low-stratigraphic and geochemically primitive mafic assemblage of komatiites and tholeiitic basalts and (2) overlying geochemically evolved calc-alkaline basalt-andesite-dacite volcanic assemblage, an example being the Abitibi greenstone belt. The two assemblages are distinguished by higher incompatible elements and light/heavy rare-earth element fractionation in the calc-alkaline assemblage. The contrast is accounted for in part by different degrees of melting of unmodified mantle lherzolite and in part by variations in the composition of the source mantle. Trace-element geochemistry studies of three Abitibi (~2.7 Ga) volcanic sequences, each approximately 16 km-thick, confirms the existence of Archaean volcanic cycles composed of lower tholeiitic and upper calc-alkalic parts (Capdevila et al. 1982). Depleted and undepleted magmas are intercalated, indicating contemporaneous coexistence of heterogeneous mantle sources. The tholeiitic basalts closely resemble modern mid-ocean ridge basalts whereas the calc-alkalic andesites and alkalic rocks are similar to modern oceanic island-arc andesites and to some modern volcanic-arc high-K rocks, respectively.

Detailed geochemical studies of Archaean granitoids (Martin 1999; Smithies 2000; Smithies and Champion 2000; Smithies et al. 2003; Champion and Smithies 1999, 2007; Martin et al. 2005; Moyen et al. 2007, 2009) suggest an overall rise in #Mg values, Ni, Cr and Sr with time (Fig. 6.10), possibly representing increased interaction between the mantle. Moyen et al. (2007) interpreted the geochemistry of Barberton TTG in terms of the depth of melting of source amphibolites, including differentiation between two sub-series – a high-Sr high-pressure low temperature sub-series of mostly leucocratic trondhjemites and a low-Sr low-pressure and high temperature sub-series including tonalite, diorite, trondhjemites and granodiorites. An evolutionary trend is observed from ~3.55 to 3.50 Ga low-Sr tonalites signifying residual plagioclase to ~3.45 Ga high-Sr trondhjemite signifying deep-level melting and thus little plagioclase in the residue, to ~3.29–3.24 Ga high-Sr tonalite and trondhjemite. The dominant tonalite-trondhjemite-granodiorite TTG series in pre-3.2 Ga terrains consists of Na-rich, high-Sr, high La/Yb and high Sr/Y compositions attributed to partial melting of eclogites. Whereas TTG display similarities to Cainozoic adakites, the latter bear evidence for stronger interaction with the mantle in terms of their high #Mg values (Martin 1999; Martin et al. 2005). The dominant pre-3.2 Ga tonalite-trondhjemite-granodiorite suite in the Pilbara Craton is replaced in post-3.0 Ga terrains by a post-kinematic high-Mg (#Mg ~ 60) diorite suite (sanukitoid) suggestive of interaction with the mantle (Smithies and Champion 2000).

Variations occur between cratons, for example a rise in K/Na about 3.3 Ga in some granitoids in the Pilbara Craton is not reflected in 3.55–3.21 Ga tonalites in the Barberton terrain which is dominated by low K/Na ratios of <0.5 (Moyen et al. 2009) (Fig. 6.9). In the Barberton terrain high εNd (CHUR) (<4‰) and low εSr (T) tonalite and granodiorite (~3.4–3.5 Ga Theespruit and Steynsdorp plutons), signifying mantle signatures, are succeeded by low εNd (CHUR) (−0.5 to −3‰) and high εSr (T) (~3.2 Ga Kaap Valley and Nelshoogte plutons) (Moyen et al. 2007, 2009).

Chapter 7
Pre-3.2 Ga Evolution and Asteroid Impacts of the Barberton Greenstone Belt, Kaapvaal Craton, South Africa

Abstract The discovery of near to 14 Archaean impact ejecta units up to 3.48 Ga-old in the Barberton Greenstone Belt, Kaapvaal Craton, South Africa, and the Pilbara Craton, Western Australia, including asteroid impact clusters about 3.25–3.22 Ga and 2.63–2.48 Ga in age, may represent discrete impact episodes or terrestrial vestiges of an extended late heavy bombardment (LHB). The interval of ~3.25–3.22 Ga-ago emerges as a major break in Archaean crustal evolution when, as indicated by a large body of field and isotopic age evidence, impacts by a cluster of large asteroid bombardment resulted in faulting, large scale uplift, intrusion of granites and an abrupt shift from crustal conditions dominated by mafic-ultramafic crust and tonalite-trondhjemite-granodiorite plutons, to semi-continental conditions represented by arenites, turbidites, conglomerate, banded iron formations and felsic volcanics. These events resulted in geotectonic patterns from pre-3.2 Ga dome-structured granite-greenstone systems such as in the Kaapvaal, Zimbabwe and Pilbara cratons, to linear accretional granite-greenstone systems such as the Superior Province in Canada, Yilgarn Craton and Pilbara Craton in Western Australia.

Keywords Asteroid impact • South Africa • Crustal evolution • Kaapvaal Craton • Zimbabwe Craton • Barberton Greenstone Belt • Onverwacht Group • Komatiites • Fig Tree Group

7.1 The Barberton Greenstone Belt, Eastern Kaapvaal Craton

The recorded evolution of the Kaapvaal Craton, spanning over 10^9 years as recorded by detailed stratigraphy and isotopic age determinations, involves a number of discrete spatial-temporal terrains (Viljoen and Viljoen 1969a, b, c; Anhaeusser et al. 1969; Anhaeusser 1973; Kroner et al. 1989, 1991a, b; Poujol et al. 1996; Eglington and Armstrong 2004; Poujol et al. 2003; Schmitz et al. 2004; Poujol 2007) (Figs. 7.1 to 7.17).

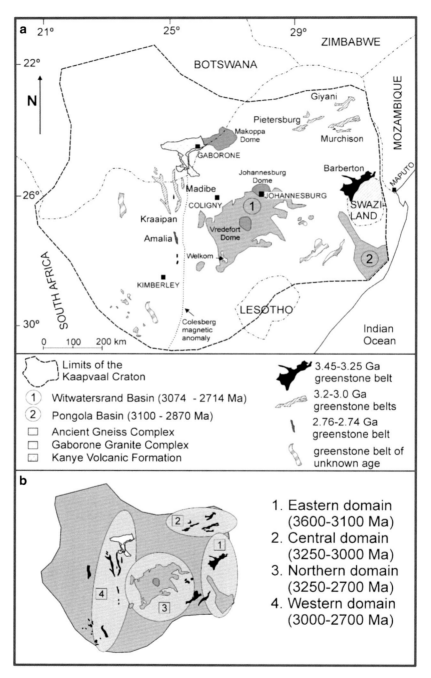

Fig. 7.1 (a) Outline of the Kaapvaal Craton, South Africa; (b) Locations of the eastern, central, northern and western domains (From Poujol et al. 2003; Elsevier, by permission)

7.1 The Barberton Greenstone Belt, Eastern Kaapvaal Craton

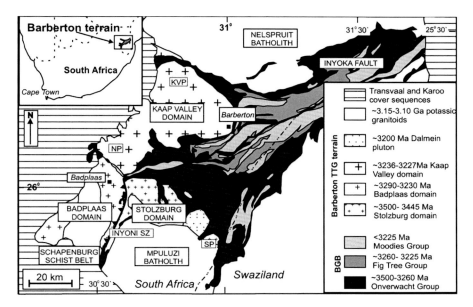

Fig. 7.2 Geological sketch map of the Barberton granitoid-greenstone terrain, showing the main lithotectonic units: *NP* Nelshoogte Pluton, *KVP* Kaap Valley Pluton, *SP* Steynsdorp Pluton. The Inyoni shear zone is marked by black line and teeth (pointing to the upper plate); the trace of the southern extensional detachment at the base of the greenstone belt is marked by solid black line (From Kister et al. 2010; Elsevier, by permission)

Structural and geophysical data allow definition of several blocks, including the Witwatersrand, Kimberley, Pieterburg and Swaziland blocks, separated by major lineaments (Fig. 7.1). Poujol (2007) defines three periods of maximum magmatic activity about ~3.55, ~3.45 and ~3.25–3.22 Ga, recognized in the Ancient Gneiss Complex of Swaziland (AGC) and the Barberton Greenstone Belt (BGB) (Fig. 7.2). The 3.55–3.22 Ga Barberton Greenstone Belt (BGB) (Fig. 7.2) and associated felsic plutons in the Eastern Transvaal and Swaziland, South Africa, along with the Pilbara craton (Chap. 8), constitutes the oldest well preserved mid-Archaean crustal fragments where primary volcanic and sedimentary features, including asteroid impact ejecta units are well retained. Allowing observations of ancient marine and terrestrial environments, with implications for the composition of the atmosphere-ocean system and early life (Viljoen and Viljoen 1969a, b; Knauth and Lowe 1978, 2003; Lowe and Knauth 1977, 1978; Lowe and Byerly 1986a, b; Lowe and Byerly 1999; Lowe and Nocita 1999; Lowe and Fisher 1999; Lowe et al. 2003; Lowe and Byerly 2007).

The Barberton Greenstone Belt (Fig. 7.2) crop out in ridges (Fig. 7.4) while the granitoids underlie valleys or plateaus. Low metamorphic grade supracrustal sequences contain ultramafic lavas – termed 'komatiites' after their type locality on the Komati River, including basaltic komatiites (high-Mg basalts) and peridotitic komatiites (Viljoen and Viljoen 1969a, b; Sun and Nesbitt 1977, 1978; Arndt et al.

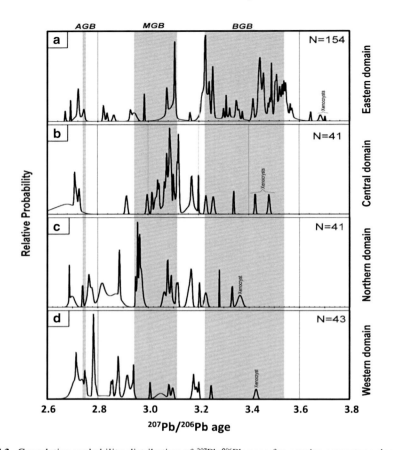

Fig. 7.3 Cumulative probability distribution of $^{207}Pb/^{206}Pb$ ages for granite-greenstone domains of the Kaapvaal Craton. The grey background tones represent the times of emplacement of the Barberton (*BGB*), Murchison (*MGB*) and Amalia (*AGB*) greenstone belts. Poujol et al. 2003; Elsevier, by permission

1997) (Figs. 7.5 and 7.6). Plutonic bodies surrounding and intruding the BGB are dominated by the plagioclase-rich tonalite-trondhjemite-granodiorite (TTG) assemblage (Glikson and Shertaon 1972; Glikson 1979a, b; Arth and Hanson 1975; Barker 1979) (Fig. 6.8, 7.7). The TTG are in turn intruded by a younger granite-monzonite-syenite suite (GMS) dominated by K-feldspar and quartz, which due to the resistance of quartz to erosion forms plateaus such as the Lochiel Plateau (Fig. 7.7b). Isotopic age determinations indicate a contemporaneity of TTG plutons and felsic volcanic units in the BGB (Appendix I) whereas the GMS association mostly postdates deformation and onset of late stage sedimentation (Lowe and Byerly 2007).

7.1 The Barberton Greenstone Belt, Eastern Kaapvaal Craton

Fig. 7.4 Two views looking northeast along the Kromberg syncline, showing the Pigge Peak ridge section with the Onverwacht Group mafic-ultramafic volcanics (ON), Fig Tree Group turbidites (FT) and Moodies Group Conglomerates (MO)

The evolution of the Barberton Greenstone Belt (BGB) can be summarized in terms of the following stages (see Appendix 1, Table 1):

Stage I (~3.7–3.5 Ga): Development of the Ancient Gneiss Complex of Swaziland, in part predating the BGB.

Stage II (~3.55–3.23 Ga): Mafic and ultramafic volcanism accompanied with accretion of co-magmatic felsic plutonic and volcanic magmatic increments (Viljoen and Viljoen 1969c), forming an early greenstone-granitoid nucleus. This stage includes four magmatic cycles (Lowe and Byerly 2007): (1) >3,547–~3,500 Ma (Steynsdorp cycle); (2) ~3,500–~3,435 Ma (Songimvelo cycle); (3) ~3,435–3,330 Ma (Kromberg cycle); (4) ~3,330–3,225 Ma (Umuduha and Kaap Valley cycle).

Fig. 7.5 Onverwacht Group volcanics: (**a**) pillowed ultramafic lava; (**b**) pillowed mafic lavas; (**c**) spinifex (quench) textured volcanic rock; (**d**) sericite schist, signifying altered felsic lava

Stage III (3,230–3,216 Ma): A major impact cluster triggering major deformation of the greenstone-granitoid nucleus, emplacement of tonalite plutons and sharp unconformities (Lowe et al. 1989, 2003).

Stage IV (~3,216–3,100 Ma): Deformation and intrusion of large granitic batholiths, uplift, large scale erosion and deposition of turbidites, BIF and felsic volcanics (Fig Tree Group) and conglomerates (Moodies Group) (Fig. 7.8)

In the BGB an early plutonic-volcanic association is represented by the ~3,510 Ma Steynsdorp pluton and felsic volcanics dated at 3,548–3,530 Ma (Kröner et al. 1992, 1996). Felsic volcanics of the Theespruit and Hoogenoeg Formations (U-Pb zircon ages 3,453–3,438 Ma) overlap the ages of the Doornhoek, Theespruit and Stolzburg plutons (3,460–3,430 Ma).

Lowe and Byerly (2007) classified the southern and central BGB in terms of 5 tectono-stratigraphic suites including both supracrustal volcanics and sediments and plutonic units juxtaposed within fault-bounded blocks:

1. Steynsdorp suite, which includes pre-3.5 Ga supracrustals and the 3,509 Ma Steynsdorp Pluton;
2. Songimvelo suite, south-central part of the BGB, including supracrustals of the Theespruit and Sandspruit Formations and ~3,445 Ma Theespruit, Stolzburg, and Doornhoek TTG plutons (Robb and Anhaeusser 1983; Kisters and Anhaeusser 1995).
3. Kromberg suite, consisting of volcanic and sedimentary rocks of the Kromberg Formation;

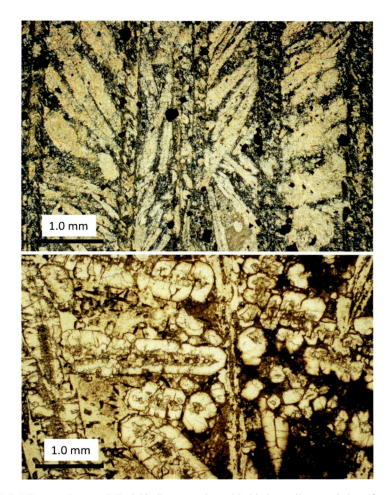

Fig. 7.6 Microscopic quench "spinifex" textures in peridotitic komatiite, consisting of tremolite and chlorite representing pseudomorphs after pyroxoene and olivine. Komati Formation, Barberton Mountain land

4. Umuduha suite, mostly supracrustals occupying the central part of the BGB;
5. Kaap Valley suite, includes supracrustals and ~3,236 Ma to ~3,225 Ma TTG plutons – the Kaap Valley, Nelshoogte, and Badplaas plutons.

Plutonic gneisses associated with the Songimvelo suite are replete with mafic and ultramafic enclaves and xenoliths correlated with the Sandspruit Formation (Viljoen and Viljoen (1969a) (Fig. 7.2), which includes thin clastic sedimentary units containing 3,540–3,521 Ma-old detrital zircons similar in age to the Theespruit Formation (Dziggel et al. 2006). The >3,547 to <3,225 Ma stratigraphy of the BGB comprises the following major groups (Viljoen and Viljoen 1969a, b; Lowe and Byerly 1999, 2007):

Fig. 7.7 (a) layered felsic-mafic agmatite within Theespruit tonalite gneiss; (b) banded and veined migmatite of the Nelspruit Migmatite, Lochiel Plateau, Barberton granite-greenstone terrain

1. Onverwacht Group (>3,547–3,260 Ma, >10 km-thick). In the southern BGB the Onverwacht Group is dominated by mafic to ultramafic volcanics intercalated by thin cherts and silicified sediments (Fig. 7.5). Along northern and western fringes of the BGB the Onverwacht Group is represented by thinner (~1,000 m) post-3.3 Ga sequence.
2. Fig Tree Group (3,260–3,225 Ma, ~1,800 m-thick) consists of felsic volcaniclastic tuffs, clastic units eroded from the underlying volcanic sequences, and banded iron formations (BIF).
3. Moodies Group (post-3,225 Ma, ~3,000 m-thick) consists of coarse-grained quartz-rich and feldspathic sandstone and chert-clast conglomerate derived by erosion of underlying granitoids and greenstone units (Fig. 7.8).

Fig. 7.8 Conglomerate of the Moodies Group displaying sheared pebbles of chert and granite

7.1.1 Onverwacht Group

~3.53 Ga Sandspruit and Theespruit Formations.
The lowermost stratigraphic units of the BGB, exposed at the Steynsdorp anticline (Fig. 7.10), consist of komatiite and basalt of the Sandspruit Formation. In both the Steynsdorp and Onverwacht anticlines lower stratigraphic units include felsic volcanics, felsic breccia, banded cherts, and mafic-ultramafic volcanics of the Theespruit Formation, the oldest U-Pb isotopic date being $3{,}538 \pm 6$ Ma. The supracrustals are faulted against older TTG gneisses dated as $3{,}538 \pm 6$ Ma.

~3.48 Ga. Komati Formation
The overlying Komati Formation consists of a ~3,500 m-thick sequence of partly pillowed quench-textured ("spinifex-textured") peridotitic komatiites, basaltic komatiites and tholeiitic basalts (Viljoen and Viljoen 1969a, b; Viljoen et al. 1983; Cloete 1999; Dann et al. 1998) (Figs. 7.5 and 7.6). Peridotitic and basaltic komatiites consist of olivine-pyroxene-chromite assemblages set in aphanitic to recrystallized meta-glass matrix. The original olivine and pyroxene are mostly altered. Olivine is replaced by pseudomorphs consisting of serpentine and magnetite. Pyroxene is commonly replaced by tremolite and chlorite. Chromites are well preserved. Only a single 5–10 cm thick intercalation of sediments and felsic tuff, dated as $3{,}481 \pm 2$ Ma (Dann et al. 1998) is known and individual pulses of lava are difficult to distinguish due to little alteration of flow tops – suggesting near-continuous eruption (Lowe and Byerly 2007).

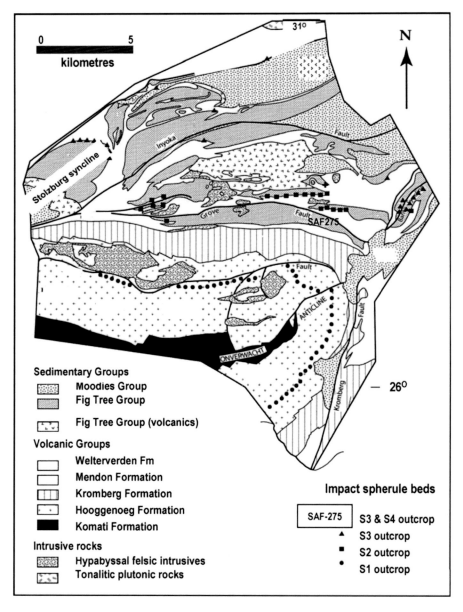

Fig. 7.9 Geological sketch map of the southwestern Barberton Greenstone Belt, east Kaapvaal Craton, South Africa, showing distribution of the S1, S2, S3 and S4 Impact spherule units (From Lowe et al. 2003; Elsevier by permission)

7.1 The Barberton Greenstone Belt, Eastern Kaapvaal Craton

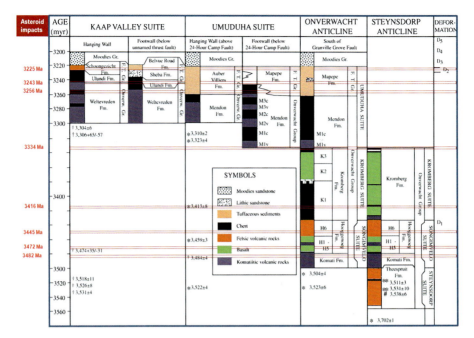

Fig. 7.10 Generalized stratigraphies of the principal tectono-stratigraphic suites in the Barberton Greenstone Belt terrain. Ages with *asterisks* (∗) below stratigraphic columns are from xenocrysts within the magmatic rocks on that block. Ages with *daggers* (†) are from detrital zircons and rock fragments within sedimentary units. Age with *number symbol* (#) is from a gneiss block within a shear zone along the southern edge of the Songimvelo Block. Ages with *double asterisks* (∗∗) are from the Theespruit Formation of the Steynsdorp suite south of the Komati Fault on the west limb of the Onverwacht Anticline (Lowe and Byerly 2007; Elsevier, by permission)

~3.47 Ga. Middle Marker

H-1. The Middle Marker forms a 1–5 m-thick silicified unit, dated as 3,472 ± 5 Ma (Armstrong et al. 1990), which caps a ~50 m-thick altered zone of the ultramafic-mafic Komati Formation. It consists of a shallow water (<100 m) wave and current-deposited assemblage of silicified ultramafic ash represented by green chert, ash particles, rip-up ash clasts, rounded carbonaceous clasts, carbonaceous sediments and accretionary lapilli (Lanier and Lowe 1982). The middle Marker correlates closely with chert and impact ejecta of the Antarctic Chert Member of the Mount Ada Basalt, central Pilbara Craton, Western Australia, dated as 3,470.1 ± 1.9 Ma-old (Byerly et al. 2002; Glikson et al. 2004) and thereby likely formed contemporaneously to an asteroid impact event (Chap. 8; Figs. 8.14, 8.15).

3,445–3,416 Ma. Hoogenoeg Formation

The Hoogenoeg Formation (Viljoen and Viljoen 1969b) consists of ~2,900–3,900 m-thick sequence of basalts, komatiite basalts, lesser amount of peridotitic basalt, dacite and cherts intercalations (Lowe and Byerly 1999, 2007). These authors classified the stratigraphy in terms of six members occupying consistent stratigraphic positions around the Onverwacht Anticline and Kromberg Syncline (Figs. 7.4, 7.9 and 7.10).

H-2. Tholeiitic basalts ~1,200–1,400 m-thick sequence of pillowed and massive tholeiitic basalt consisting of ~10–50 m-thick flow units and capped by chert (Viljoen and Viljoen 1969b; Williams and Furnell 1979);

H-3. Komatiite and tholeiitic basalts ~220–380 m thick, including pillowed and variolitic tholeiitic basalt, with komatiite basalt at the top, which consists of a massive ~150 m-thick ultramafic unit (Rosentuin Ultramafic Body, Viljoen and Viljoen 1969b) overlain by silicified volcaniclastic, accretional lapilli, carbonaceous sediments, altered ash. Trace element patterns in the ash, including high Cr, suggest derivation from ultramafic lava (Lowe 1999a, b).

H-4. ~250–350 m thick komatiite basalt including a 10–20 m thick unit consisting of very coarse-grained quench-textured pyroxene, overlain by intercalated non-quench textured pillowed basalt. A silicified intercalation includes a 0–6 m-thick silicified tuff, volcaniclastic, carbonaceous chert and an impact ejecta unit S-1 dated from detrital zircons as 3,470 ± 2 Ma (Byerly et al. 2002).

H-5. ~390 m thick sequence of massive to pillowed high-Mg and tholeiitic basalt. The pillowed basalt display radial pipe vesicles, variolite textures <3 cm-large and chilled selvages. As in Pilbara pillow basalts (Fig. 8.6), varioles (osceli) decrease in size, and become increasingly well-defined from pillow cores toward chilled pillow margins, indicating outward boiling of volatile during cooling and likely liquid immiscibility (Ferguson and Currie 1972). Intercalations of chert include basal black chert, volcaniclastic sediments and accretional lapilli.

H-6. A 1,000–2,500 m-thick sequence of felsic and volcaniclastic volcanics including quartz-bearing dacite (Viljoen and Viljoen 1969b) dated as 3,445 ± 4 Ma (de Wit et al. 1987) capped by volcanic breccia, conglomerate and arenites. An intrusive origin of the dacite below the H-4 and H-5 volcanic units is indicated by basaltic enclaves derived from lower stratigraphic levels (Lowe and Byerly 2007). The massive dacite sill or dome is flanked laterally by felsic volcaniclastic breccia and conglomerate dated as 3,445 ± 6 Ma and 3,438 ± 12 Ma (de Wit et al. 1987) and silicified turbidites (Viljoen and Viljoen 1969b; Lowe and Knauth 1977). Possible co-magmatic relations between the H-6 intrusive dacite and 3,445 Ma-old TTG plutons south of the BGB were considered by de Wit et al. 1987. Knauth and Lowe (2003) show that, whereas Hoogenoeg Formation sediments display part equilibration of $\delta^{18}O$ compositions with those of associated volcanics, stratigraphically higher sediments retain non-igneous compositions, indicating lesser degrees of hydrothermal alteration.

3,416–3,334 Ma. Kromberg Formation

The base of the Kromberg Formation (>3,416 ± 7 Ma, Kröner et al. 1991a, b) consists of 100–200 m serpentine-altered ultramafics (Viljoen and Viljoen 1969b) overlain by about 1,700 m of massive to pillowed basalt and komatiite, mafic lapilli tuff and banded chert whose top consists of a stratigraphically consistent black chert – the

Footbridge Chert overlain by the volcanic Mendon Formation. The Kromberg Formation includes the following units (Lowe and Byerly 2007).

K-1. Buck Reef Chert. A 150–350 m-thick unit of chert overlying dacite volcaniclastic arenites and shallow water carbonaceous sediment and evaporite of the Hoogenoeg Formation (Lowe and Fisher 1999). The basal section consists of silicified evaporite, a lower section consists of banded chert and ferruginous chert and shallow water sediments. The cherts include carbonaceous components representing microbial activity (Tice and Lowe 2006). The sediments are locally intercalated with basaltic volcanics and are locally underlain by peridotitic komatiite and accretional lapilli (Lowe and Byerly 2007). The Black Reef Chert is locally eroded at the top and is intersected by mafic dykes and sills, representing feeders for overlying volcanic units.

K-2. Up to 1,000 m-thick sequence of mafic lapilli tuff. Mafic coarse-grained tuff and lapilli, including a basal unit of 10–100 m-thick mafic to komatiitic tuff, thin komatiite flows and tuffaceous carbonaceous shale overlain by 300–1,000 m of mafic to ultramafic Cr-rich lapilli, including 0.5–4 cm large lapilli, chert fragments and volcanic fragments, intercalated with mafic flows (Ransom et al. 1999). Overlying 50–100 m-thick units feature finer grained (<1 cm) lapilli and ash deposited in wave rippled shallow water environment and extensively altered by chlorite, tremolite, iron-rich dolomite, and ankerite.

K-3. K2 is overlain by 500–600 m thick sequence of silicified pillow basalt with the upper 350 m consisting of tholeiitic basalt and intercalations of komatiite, tuff and pillow breccia, overlain by a unit of 15–25 m of black and white banded chert (Footbridge Chert). Felsic tuff from the Footbridge Chert yielded zircon ages of <3,334±3 Ma (Byerly et al. 1996).

3,330–3,258 Ma. Mendon Formation (Lowe and Byerly 1999) (Fig. 7.10). Multiple volcanic cycles composed of ultramafic flows overlain by chert (Byerly 1999). Locally the formation consists of 300 m-thick peridotite komatiite flow capped by 20–30 m of graded silicified ultramafic ash, accretional lapilli and carbonaceous sediments (Msauli Chert). Lowe (1999b) interprets the Msauli Chert in terms of eruption of komatiitic ash into marine environment where quiet water conditions alternate with current activity. Locally the Mendon Formation is overlain by 20–50 m of black banded and ferruginous chert underlying the Fig Tree Group.

3,258–3,227 Ma. Fig Tree Group (Fig. 7.10) The Mapepe Formation of the Fig Tree Group consists of 700 m of predominantly fine-grained felsic tuff, shale, chert, arenite, grit and chert-clast conglomerate. A northern section includes the Ulundi Formation, Sheba Formation, Belvue Road and Schoongezicht Formations, which total ~1,500 m. The Ulundi Formation (20–50 m) consists of carbonaceous shale, ferruginous chert and iron-rich sediments, including impact unit (S3), the Sheba Formation (500–1,000 m) consists of fine- to medium-grained greywacke turbidite and the Belvue Road Formation (200–500 m) consists of mudstone and thin, fine-grained turbidite arenite and dacite volcani-clastics.

South of the Inyoka Fault the Group consists of the Mapepe and Auber Villiers Formations (Fig. 7.10) .At its base and mid-level the Mapepe Formation includes asteroid impact ejecta units (S2, S3, S4) (Lowe and Byerly 1986b; Lowe et al. 1989, 2003; Shukolyukov et al. 2000; Kyte et al. 2003), overlain by clastic sediments, dacite to rhyodacite volcani-clastics and Jaspilite. The basal impact unit (S2) consists of impact condensation spherules (microkrystites, Glass and Burns 1988) and related ejecta (Lowe et al. 2003) overlain by hematite-rich sediments (Manzimnyama Jaspilite), fine-grained felsic tuffs dated as 3,258 ± 3 Ma–3,227 ± 4 Ma (Kröner et al. 1991a, b; Byerly et al. 1996), delta clastics, chert, jasper and barite – locally forming barite deposits (Heinrich and Reimer 1977). These units are overlain by 1,500–2,000 m of dacitic tuff; coarse volcaniclastic, arenite, conglomerate, breccia, gritstone and chert-clast conglomerate dated as 3,253 ± 2 Ma (Byerly et al. 1996) and 3,256 ± 4 Ma (Kröner et al. 1991a, b). Dacite located at high stratigraphic levels of the Fig Tree Group yields a zircon age of 3,225 ± 3 Ma (Kröner et al. 1991a, b).

The Fig Tree Group truncates underlying Onverwacht Group and Mendon Group through major unconformities (Figs. 7.9 and 7.10). Little or no detrital K-feldspar is reported from sediments which precede deposition of the Fig Tree Group (Lowe and Byerly 2007), suggesting differentiated granitoids were not exposed during 3.55–3.258 Ga times.

3,224–3,109 Ma. Moodies Group

The Barberton Greenstone Belt is capped by a 3,500 m-thick sequence of lithic, feldspathic, and quartz-rich arenite, conglomerate and siltstone defined as the Moodies Group, unconformably and locally conformably overlying uplifted older Archaean crust and representing deposition in fluvial to tidal marine environments in foreland and intermontane basins (Eriksson 1978, 1979, 1980; Heubeck and Lowe 1999) (Fig. 7.8). North of the Inyoka Fault the Moodies Group (>2,000 m) includes K-feldspar-bearing granitoid pebbles, indicating erosion of late-stage plutons associated with the BGB. The older age limit of the Moodies Group is inferred from the age of the youngest granitoid clasts (3,224 ± 6 Ma, Tegtmeyer and Kröner 1987) whereas the upper age limit is defined by post-tectonic intrusions, such as the Salisbury Kop Pluton (3,109 + 10/−8 Ma) which truncates tightly folded Moodies strata (Kamo and Davis 1994), and a felsic dike intrusive into Moodies Group was dated as 3,207 ± 2 Ma (Heubeck and Lowe 1994).

7.2 Evolution of the Zimbabwe Craton

The Zimbabwe craton comprises several distinct terrains, including (1) the central ~3.5–2.95 Ga Tokwe gneiss terrain containing inliers of greenstone belts, (2) unconformably overlying ~2.9–2.8 Ga mafic and felsic volcanic rocks and conglomerates, and a ~3.0–2.7 Ga sequence of sandstone, shale, and limestone; (3) Greenstone belts flanking the central Tokwe terrain, including ~2.7 Ga northwestern greenstone belts consisting of calc-alkaline volcanics and intercalated sediments intruded by syn-volcanic plutons, and ~2.7 Ga southeastern greenstone belts consisting of thick piles of tholeiitic basalts overlying ultramafic lavas (Kusky 1998) (Chap. 3). Oldest detrital zircons yield ages up to ~3.8 Ga (Jelsma and Dirk 2000). Early Archaean zircon xenocrysts within Late Archaean volcanic rocks are interpreted by Wilson et al. (1995) in terms of their deposition on early Archaean granitoid crust. The oldest components of the Tokwe gneiss terrain include tonalite, granodiorite and migmatite yielding U/Pb and Pb/Pb ages of ~3.5–2.95 Ga (Kusky 1998), including the ~3.5 Ga gneisses and greenstones referred to as Tokwe segment (Wilson 1973, 1979; Wilson et al. 1990) (Fig. 3.1). U-Pb single zircon ages of early components of the Tokwe gneiss terrain yielded 3,565±21 Ma whereas other components are dated as 3,455±2 Ma and 3,456±6 Ma from the Tokwe gneiss terrain (Horstwood et al. 1999). Younger intrusions within the gneiss terrain yielded ages of ~3,350 Ma, indicating an upper age limit high grade metamorphism.

The ~3.5 Ga greenstones, denoted Sebakwian Group, are intimately intercalated with the gneisses (Fig. 3.1), interpreted by Stowe (1973) in terms of imbricate interthrusting of mafic volcanic layers with older gneisses, subsequently intruded by younger granites. Examples are ultramafic rocks, iron formation, quartzite, and mica schist in the Mashaba area in-folded with ancient gneissic rocks (Wilson 1979). By contrast Tsomondo et al. (1992) regarded the gneiss-greenstone complex as autochthonous. In other areas the Tokwe gneisses are unconformably overlain by a volcanic-sedimentary association referred to as Lower Bulawayan Group (Blenkinsop et al. 1997). Late Archaean volcanic activity lasted between ~2.9–2.65 Ga, including at least three ultramafic-mafic to felsic to late K-granite magmatic cycles (Wilson et al. 1995). In the Belingwe greenstone belt these supracrustals include mafic, ultramafic, intermediate, and felsic volcanic rocks, pyroclastic deposits and sedimentary rocks (Bickle et al. 1975), dated by U-Pb isotopic studies as 2,904±9 Ma and 2,831±6 Ma (Wilson et al. 1995). Upper age limits to this belt are defined by Pb-Pb ages of 2,833±43 Ma of the Chingezi tonalite (Taylor et al. 1991). Lower Bulawayan rocks are also present in the Midlands, Filabusi, Antelope–Lower Gwanda, Shangani, Bubi and Gweru-Mvuma greenstone belts, including post-3.09 Ga, pre-2.86 Ga shallow-water shelf arenites and banded ironstones and deep water shale, chert, BIF, carbonate and associated mafic-ultramafic volcanics (Fedo et al. 1996). The unconformably above post-2,831±6 Ma ~600 m-thick Manjeri Formation sediments (Wilson et al. 1995), which overlap the Tokwe basement gneisses and Sebakwian greenstones, is overlain by conglomerates, arenites and carbonate which grade upward into chert, argillite, greywacke and BIF (Bickle and Nisbet 1993). Kusky (1998) classified the Bulawayan Group into contemporaneous northern and southern volcanic belts possessing distinct lithologies, geochemistry and field relations:

Northern volcanic belts: The belt is located northwest of the Tokwe gneiss terrain, comprising the Bulawayan, Harare-Shamva, and other greenstone belts consisting of lower volcanic units succeeded by calc-alkaline basalt-andesite-dacite-rhyolite volcanic suite intruded by syn-volcanic plutons, dated as 2.68–2.70 Ga (Wilson et al. 1995; Wilson 1979). U-Pb, Pb-Pb, and Sm-Nd isotopic studies of the Harare-Shamva supracrustals and associated granites indicate crustal contamination of 2.72–2.67 Ga volcanics followed by deformation about 2.66 Ga and post-tectonic granitoids 2.65–2.60 Ga (Jelsma et al. 1996).

Southern volcanic belts: The belt includes the Belingwe greenstone belt, consisting of tholeiitic mafic-ultramafic volcanics which overlie gneisses and shallow-water sedimentary sequences southwest of the Tokwe gneiss terrain. In the Belingwe greenstone belt high-magnesium basalts, komatiites, and their intrusive rocks of the Reliance Formation and Zeederbergs Formation (2,692 ± 9 Ma, Chauvel et al. 1993) overlie sediments of the Manjeri Formation through allochtonous contacts (Bickle and Nisbet 1993). Komatiites of the Reliance Formation are regarded as formed on an oceanic plateau (Kusky and Kidd 1992) and are geochemically similar to intra-plate basalts (McDonough and Ireland 1993). Oceanic-like basalts of the Tati belt are overlain by andesite and felsic volcanics intruded by syn-tectonic granitoids (Key et al. 1976; Aldiss 1991).

Shamvaian Group: The group forms the uppermost Archaean supracrustals, consisting of terrestrial clastic sediments which unconformably overlie 2.7 Ga greenstone belts. The Shamvaian Group is 2 km-thick in the Bindura-Shamva greenstone belt where it consists of basal conglomerates which grade into thick arenite sequence (Jelsma et al. 1996). Felsic volcanics associated with the Shamvaian Group are intruded by circa ~2.6–2.57 Ga Chilimanzi Suite K-rich granitoids (Wilson et al. 1995; Blenkinsop et al. 1997). The Cheshire Formation which overlies the Belingwe belt consists of conglomerate, sandstone, siltstone, argillite, limestone, and minor banded iron formation (Bickle and Nisbet 1993). These events overlap and are likely related to the 2.68–2.58 Ga tectono-magmatic events in the Limpopo belt at the southern margin (Treloar and Blenkinsop 1995). Stabilization of the Zimbabwe Craton was near-complete with the emplacement of the Great Dyke (Fig. 3.1a) at 2,575.4 ± 0.7 Ma (Oberthur et al. 2002).

Jelsma and Dirk (2000) sum up the Late Archaean 2.7–2.6 Ga history of the Zimbabwe craton in terms of modern-like plate tectonics, involving diachronous accretion of diverse crustal fragments including oceanic and back arc mafic crust, volcanic arc felsic crust, continental crust, and related sedimentary sequences, including concomitant diapirism, strike-slip faulting, magma intrusion, and sedimentation. The extent to which modern tectonic processes apply to Archaean terrains is considered in Chap. 12.

7.3 Pre-3.2 Ga Asteroid Impact Units of the Kaapvaal Craton

Discoveries of impact spherule and ejecta units in the Archaean Barberton Greenstone Belt (BGB) (Lowe and Byerly 1986a, b, 2010; Lowe et al. 1989, 2003; Lowe 2013) include the following impact ejecta units (Figs. 7.11 to 7.17):

~3,482 Ma – Komati Formation, Onverwacht Group (BGB)
~3,472 Ma – S1 spherule unit – Hoogenoeg Formation, Onverwacht Group (BGB)
~3,445 Ma – Hoogenoeg Formation, Onverwacht Group (BGB)
~3,416 Ma – top Hoogenoeg Formation, Onverwacht Group (BGB)

7.3 Pre-3.2 Ga Asteroid Impact Units of the Kaapvaal Craton

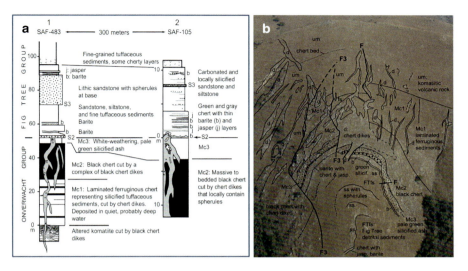

Fig. 7.11 Barberton impacts spherule-bearing chert dykes: (**a**) Stratigraphic sections of uppermost Onverwacht and lowermost Fig Tree Groups in the western part of the Barite Valley. Note differences in scales of columns. (**b**) Aerial photo, approximately perpendicular to stratification, of the type 1 chert dike complex A. (**a**) Photo without contacts. Scale is approximate because of perspective. (**b**) Photo with geologic contacts showing the restriction of chert dikes to units below the Fig Tree Group. One tongue of black chert extends upward through spherule bed S2 (*black dots*) and overlying green silicified ash units, either through faulting or early post diking foundering of dense barite-rich sediments into the still-soft dike complex. Several later faults, including F3, crosscut the dike complex and extend upward into the Fig Tree Group. Symbols: um—altered ultramafic volcanic rocks; Mc1, Mc2, Mc3 (*stippled*)—subdivisions of the Mendon sedimentary section; b—barite; ss—Fig Tree sandstone; d—type 1 chert dikes (From Lowe 2013; Geological Society of America, by permission)

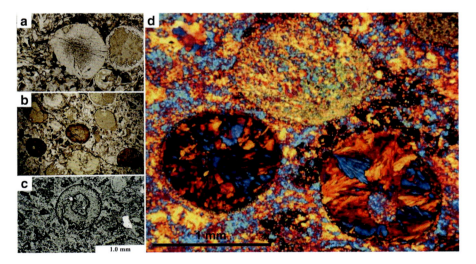

Fig. 7.12 (**a–c**) Barberton ~3,256 Ma S2 microkrystite spherules displaying quench textures and central to offset vesicles; (**d**) S3 spherules displaying inward radiating chlorite crystallite fans in polarized light (Courtesy G. Byerly)

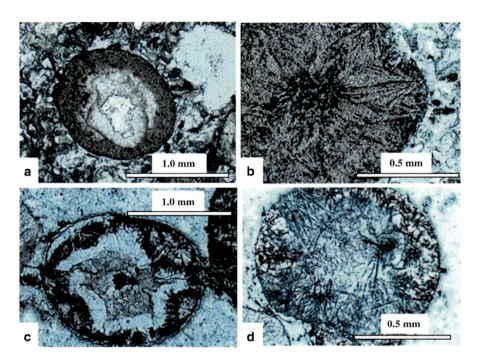

Fig. 7.13 Barberton 3,243 ± 4 Ma S3 microkrystite spherules: Plane polarized microphotographs of microkrystite spherules: (**a**) cryptocrystalline chlorite-mantled spherule displaying an offset central vesicle consisting of chlorite–quartz aggregates. Opaque oxide grains within the chlorite include Ni-rich chromite and sulphide. The spherule is set in groundmass consisting of chloritized to silicified mafic volcanic fragments; (**b**) chlorite dominated spherule displaying inward-radiating crystal fans interpreted as pseudomorphs after original quench structures; (**c**) An internally zoned spherule showing inward radiating chlorite aggregates, a median silica-dominated zone and an inner chlorite–quartz core; (**d**) chlorite spherule dominated by Interlocking inward-fanning needle-like pseudomorphs after pyroxene and olivine (From Glikson 2007; Elsevier by permission)

~3,334 Ma – top Kromberg Formation, Onverwacht Group (BGB)
~3,256 Ma – S2 spherule unit – Base Mapepe Formation, Fig Tree Group (BGB)
~3,243 Ma – S3 spherule unit – Base Sheba Formation, Fig Tree Group (BGB)
~3,225 Ma – Base Belvue Formation, Fig Tree Group (BGB)
~3,225 Ma – Schoengezicht Formation, Fig Tree Group (BGB)

Detailed accounts presented to date pertain to the ~3,472 Ma, ~3,256 Ma, ~3,243 Ma and ~3,225 Ma impact ejecta units:

~3,482 Ma: The impact spherules unit (Lowe and Byerly 2010) is located in sediments intercalated within mafic-ultramafic volcanic sequence of the Komati Formation stratigraphically near a tuff horizon dated as 3,482 ± 2 Ma (Dann 2000).

~3,472 Ma: The S1 spherule unit, located within a 30–300 cm-thick sericite-bearing and black chert intercalation exposed along ~25 km strike distance within the felsic

7.3 Pre-3.2 Ga Asteroid Impact Units of the Kaapvaal Craton

Fig. 7.14 (**a**) Ni-rich chromites within Barberton S3 microkrystite; (**b**) Microkrystite spherule containing quench skeletal Ni-chromite. Spherule on lower left shows an off-center vesicle typical of microkrystite spherules (plane-polarized light) (From Byerly and Lowe 1994; Elsevier, by permission)

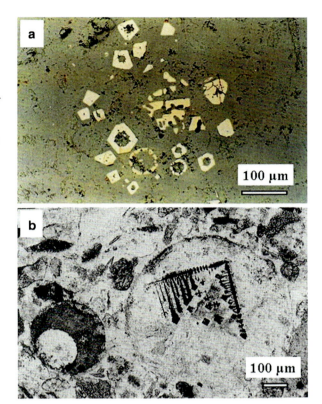

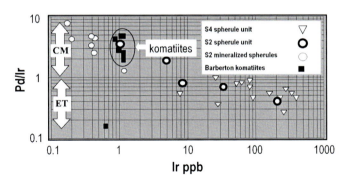

Fig. 7.15 Pd/Ir vs Ir plots for Barberton spherule units S4 and S2 compared to komatiites. The data allow a clear distinction between the high Ir low Pd/Ir field of the spherules (*ET* extraterrestrial) and the low-Ir high Pd/Ir field of komatiites (*CM* crust–mantle) (From Glikson 2008; Elsevier by permission)

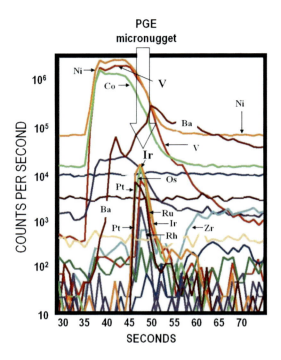

Fig. 7.16 LA-ICPMS spot analysis (counts per second vs. time in seconds) of a Ni–chromite in sample SA306-1, showing a PGE micro-nugget (From Glikson 2007; Elsevier, by permission)

volcanic Hoogenoeg Formation, and denoted S1 (Lowe et al. 2003), was dated by U-Pb zircon as 3,470 ± 3 Ma (Byerly et al. 2002). S1 constitutes a 10–35 cm-thick unit consisting of two distinct graded spherule units with 10–30 % spherules and including chert and locally volcanic fragments. The spherule units, consisting of 0.3–0.6 mm-size spherical to ovoid spherules, are intercalated with chert and tuff, locally overly chert and mafic volcanics unconformably and grade upwards into cross laminated siltstone (Lowe et al. 2003). Spherule compositions range from microcrystalline quartz to, less commonly, sericite-dominated, and are commonly rimmed by microcrystalline rutile and/or radial microlitic textures, likely representing quench effects. Internal cavities filled with quartz and sericite occur. Broken segments of spherules are common. Angular fragments consist of volcanic fragments and of quartz-sericite-rutile assemblages similar to those of the spherules. Iridium levels vary up to 3 ppb, only somewhat above those of underlying komatiites with 2.1 ppb Ir. Lowe et al. (2003) interpret the depositional setting of the host chert in terms of moderately deep below wave base environment. S1 is correlated with 3,470.1 ± 1.9 Ma microkrystite spherule units the Antarctic Chert member, North pole dome, Pilbara Craton (Byerly et al. 2002) (Figs. 8.14 and 8.15).

~3,256 Ma. The S2 spherule unit (Lowe et al. 2003) (Figs. 7.9, 7.10 and 7.12), is located between black chert at the top of the Mendon Formation of the Onverwacht mafic-ultramafic volcanic sequence and basal volcani-clastic, chert and ferruginous sediments of the overlying Mapepe Formation of the Fig Tree Group. A zircon U-Pb

7.3 Pre-3.2 Ga Asteroid Impact Units of the Kaapvaal Craton 93

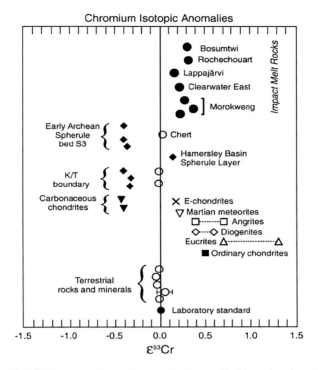

Fig. 7.17 εCr (^{53}Cr/^{52}Cr) values of several meteorite classes, K–T boundary impactites, BGB-S4 impactites and sediments. *Open circles* – terrestrial rocks; *all other symbols* – meteoritic and impact spherules compositions. Note the correspondence between impactite εCr range and carbonaceous chondrites [εCr = 10^4{(^{53}Cr/^{52}Cr(m) − (^{53}Cr/^{52}Cr(t)}/{^{53}Cr/^{52}Cr(t)}: *m* meteoritic, *t* terrestrial] (From Shukolyukov et al. 2000; permission requested from Springer)

age of 3,258 ± 3 Ma was determined on tuff located 20 m above S2 (Byerly et al. 1996). The unit ranges between 20 and 310 cm thickness consists of 0.15–2.5 mm-size spherules of pure sericite to nearly pure silica composition accompanied with microcrystalline rutile showing microlitic quench textures. The units contain about 10–50 % spherules associated with silt-size to cobble-size volcanic and sedimentary fragments derived from subjacent strata (Lowe et al. 2003). S2 includes two graded sub-layers including lithic fragments, the lower consisting of 30–50 % spherules and the upper of ~10 % spherules and showing current bedding, and capped by chert and ferruginous representing silicification of fine grained sediments. Iridium levels within S2 spherules are in the range of 0–3.9 ppb and are in part accounted for by abundant komatiite debris. An extraterrestrial origin of anomalous PGE levels is corroborated by Cr isotopic studies indicating ^{53}Cr/^{52}Cr ratios corresponding to carbonaceous chondrites (Shukolyukov et al. 2000; Kyte et al. 2003) (Fig. 7.17)

The S2 spherules (Fig. 7.12) were deposited in a turbulent waves and current-dominated environment, including locally derived detritus eroded from underlying

Onverwacht cherts and komatiitic volcanic units. This is contrasted with underlying and overlying sediments deposited under quiet, low-energy conditions, which suggests abrupt effects of seismicity and tsunami disturbances affecting otherwise pelagic below-wave base environments and associated with deposition of impact debris. In some areas the spherule unit is underlain by komatiite-clast breccia scoured into underlying chert and volcanics of the underlying Mendon Formation of the Onverwacht Group.

Lowe (2013) investigated the structure of chert dykes and veins which extend up to about 100 m into formations below the S2 spherule unit, likely representing the effects of seismic events and tsunami pressures associated with the impact/s (Fig. 7.11). Four types of veins were recognized:

Type I. irregular dykes up to 8 m wide extending downward by up to 100 m, regarded as younger than Type II veins and representing liquefaction of sea bed sediments and spherule units and followed by tsunami effects;

Type II. Small vertical veins mostly <1 m wide restricted to the lower half of the Mendon chert section and representing the earliest seismic effects;

Type III. Small cross-cutting veins mostly <50 cm wide filled with silica;

Type IV. Small irregular to bedding-parallel and irregular veins mostly <10 cm wide filled with silica. Filling of the dykes involved foundering and injection of dense barite sediments into underlying fractures under high tsunami hydrodynamic pressures. The author remarks on the abrupt change from ~300 Ma of mafic-ultramafic volcanism represented by the Onverwacht Group to the orogenic clastic sedimentation and associated felsic volcanism of the Fig Tree Group, evolved over the next ~100 Ma into the Kaapvaal Craton.

~3,243 ± 6 Ma. The S3 spherules (Figs. 7.12, 7.13, 7.14) is intercalated with tuffaceous and terrigenous sediments about the middle of the Mapepe Formation of the Fig Tree Group, in some areas directly overlying komatiite and chert of the Onverwacht Group and underlying black cherts of the Ulundi Formation (Lowe et al. 1989, 2003; Byerly and Lowe 1994). An underlying dacite tuff yielded a U-Pb zircon age of 3,243 ± 6 Ma (Kröner et al. 1991a, b). S3 is about 30–35 cm thick where it displays grading and is up to 2–3 m thick where it may have been accumulated by secondary currents. Well preserved spherules are mostly siliceous whereas chrome mica-rich spherules tend to be deformed. Centrally off-centered circular regions mostly filled-in by quartz represent original aerodynamic cavities of the microkrystites. Cr-mica may form microcrystalline quench textures radiating inward from spherule margins (Fig. 7.13). Chlorite and rutile are common components of spherules. The type section termed Jay Chert consists of 78 cm of spherules intercalated with silicified sandstone, mudstone, and intra-formational conglomerate located 100–150 m above the base of the Mapepe Formation (Lowe and Nocita 1999). Underground exposures of S3 occur in the Sheba Gold, Princeton Tunnel and Agnes Mines. Extreme iridium levels of 426–519 ppb occur in the Jay Chert spherule unit where spherules form ~10–50 % of the rock. Ir levels of 104–725 ppb are reported from in spherules from the Sheba Mine (Fig. 7.15). Cr isotopic analyses of S3 spherules indicate meteoritic origin of the Cr, suggesting a CV-type carbonaceous chondrite-type projectile (Kyte et al. 2003).

7.3 Pre-3.2 Ga Asteroid Impact Units of the Kaapvaal Craton

Spherule sizes may reach the unusual size of 4 mm in diameter (Byerly and Lowe 1994). Ni-rich spinels, including dendritic and octahedral chromites (Fig. 7.14), constitute a relic primary phase within S3 spherules (Byerly and Lowe 1994; Glikson 2007). These spherules display pseudomorphs after olivine, pyroxene and plagioclase, edge-centered fibro-radial crystals representing devitrification textures (Fig. 7.13). Byerly and Lowe (1994) distinguished two types of spinel:

1. chalcophile spinels with low MgO, Al2O3, and TiO2;
2. lithophile spinel with up to 5 % Al2O3 and MgO.

Energy dispersive spectrometry, laser ablation inductively coupled plasma mass spectrometry (LA-ICPMS) track analyses of chlorite-dominated quench-textured microkrystite spherules and LA-ICPMS spot analyses of intra-spherule Ni-rich skeletal quench chromites from the $3{,}243 \pm 4$ Ma Barberton S3 impact fallout unit reveal fractionated siderophile and PGE trace element patterns corresponding to chondrite-contaminated komatiite/basalt compositions (Glikson 2007). The chlorites, interpreted as altered glass, contain sharp siderophile elements and PGE spikes inherited from decomposed metal and Ni-rich chromite particles. LA-ICPMS spot analysis identifies PGE-rich micro-nuggets in Ni–chromites (Ir $\sim$12–100 ppm, Os $\sim$9–86 ppm, Ru $\sim$5–43 ppm) and lower levels of the volatile PGEs (Rh $\sim$1–11 ppm, Pd $\sim$0.68–0.96 ppm) (Fig. 7.16). Previously reported PGE anomalies in the order of hundreds of ppb in some Barberton microkrystite spherules (Fig. 7.15) are accounted for in terms of disintegration of PGE-rich micro-nuggets. Replacement of the Ni-chromites by sulphide masks primary chondritic patterns and condensation element distribution effects. High refractory/volatile PGE ratios pertain to both the chlorites and the Ni-rich chromites, consistent with similar compositional relations in microkrystite spherules from other impact fallout units in the Barberton Greenstone Belt and the Pilbara Craton, Western Australia. The near-consistent low Pt/Re and high V/Cr and V/Sc ratios in chlorite of the spherules, relative to komatiites, are suggestive of selective atmospheric condensation of the spherules which favored the relatively more refractory Re and V. Selective condensation may also be supported by depletion in the volatile Yb relative to Sm. Ni–Cr relationships allow estimates of the proportion of precursor crustal and meteoritic components of the spherules. Mass balance calculations based on the iridium flux allow estimates of the order of magnitude of the diameter of the chondritic projectile.

~3,225 Ma. The S4 spherule unit is known from a single outcrop within the Jay Creek Chert at 6.5 m stratigraphically above S3, cropping out as a 15 cm-thick current-bedded arenite mixed with terrigenous debris over a 1 m-long distance along strike beneath sandstone and conglomerate. The S4 spherules are 0.2–1.6 mm in diameter and in part display amalgamated textures indicating mergence in the liquid state. They contain up to 25–30 % Cr-rich rutile, Cr-rich chlorite, apatite, pyrite and chalcopyrite. Iridium levels reach 250–350 ppb and Os/Ir and Pt/Ir ratios are akin to those of CI chondrites (Kyte et al. 1992). Cr isotopic compositions of S4 are akin to those of carbonaceous chondrite, with $^{53}Cr/^{52}Cr$ ratios of 20.32 ± 0.06 and 20.26 ± 0.11 (Shukolyukov et al. 2000).

Diagnostic criteria allowing the identification of the asteroid impact origin of the spherule-bearing ejecta layers include:

1. PGE chondrite-normalized patterns displaying marked depletion in the volatile species (Pd, Au) relative to refractory species (Ir, Pt), distinct from terrestrial PGE profiles, excepting depleted mantle harzburgites (Fig. 7.15).
2. Quench-textured and octahedral Ni-chromites (NiO<23 %) with high Co, Zn and V, unknown in terrestrial chromites (Byerly and Lowe 1994), and which contain PGE nano-nuggets of compositions distinct from terrestrial PGE nuggets (Fig. 7.16).
3. $^{53}Cr/^{52}Cr$ isotopic indices ($\varepsilon^{53}Cr = -0.32$) which correspond to values of carbonaceous chondrites and values of K-T boundary impact fallout deposits, but distinct from terrestrial values (Shukolyukov et al. 2000) (Fig. 7.17);
4. Diagnostic textural features, including inward-radiating quench textures and offset central leucocratic vesicles, as defined by B.M. Simonson (1992) (Figs. 7.12 and 7.13).

Mass balance calculations based on Ir and Cr levels and thermodynamic-based correlations of spherule sizes of 1–4 mm-diameter (O'Keefe and Aherns 1982; Melosh and Vickery 1991), suggest impact by asteroids on the order of 30–50 km-diameter (Lowe et al. 1989; Byerly and Lowe 1994), scaled to terrestrial impact basins some 300–800 km in diameter. The Fe-Mg-rich composition of the spherules and the absence, in most instances, of shocked quartz in the units suggest the impact basins formed in simatic/oceanic regions of the Archaean Earth, which from geochemical and isotopic evidence (McCulloch and Bennett 1994) occupied over 90 % of the Earth surface before about 3.0 Ga. The stratigraphic location of the Barberton S2–S4 spherule units immediately above the >12 km thick mafic-ultramafic volcanic sequence of the Onverwacht Group and at the base of a turbidite/felsic volcanic sequence of the Fig Tree Group, which contains granitic detritus, hints at the onset of fundamentally different crustal regime about 3.24–3.227 Ga (Lowe et al. 1989; Glikson 2001, 2013; Glikson and Vickers 2007, 2010). Estimates of asteroid impact incidence rates and modeling of the tectonic and magmatic consequences of mega-impacts on thin thermally active oceanic crust (Glikson and Vickers 2007, 2010) suggest such impact clusters constituted critical factors in Archaean crustal evolution.

U-Pb and Lu-Hf isotope studies by Zeh et al. (2009) using laser ablation-sector field-inductively coupled plasma-mass spectrometry on zircon grains from 37 granitoid samples from the Kalahari Craton (including the Barberton, Murchison, Limpopo, Francistown cratons), indicate major crust forming events at 3.23 Ga, 2.9 Ga and 2.65–2.7 Ga. Whole-rock Sm–Nd isotope systematics of 79 Archaean granitoids by Schoene et al. (2009) indicate offsets in εNd values for 3.2–3.3 Ga granitoids consistent with significant crustal growth at this stage. A concentration of felsic plutonic activity about ~3.3–3.2 Ga is also indicated by U-Pb zircon ages of the Nelshoogte pluton (3,212±2 Ma and 3,236±1 Ma) and Kaap Valley pluton (3,229–3,223 Ma) (Appendix I) ages. The significance of the overlaps between the ~3.24–3.227 Ga Barberton impact cluster and these magmatic events requires further investigation.

Chapter 8
Evolution and Pre-3.2 Ga Asteroid Impact Clusters: Pilbara Craton, Western Australia

Abstract Volcanic and sedimentary sequences of the Pilbara Craton, Western Australia, display remarkable stratigraphic and isotopic age correlations with the Barberton Greenstone Belt (BGB), Kaapvaal Craton. However, peridotitic komatiites are less common in the Pilbara and intrusive batholiths include a lesser tonalite component than in eastern Kaapvaal plutons. Asteroid impact units in the Pilbara Craton include a ~3.47 Ga multiple spherule ejecta units precisely correlated with an impact ejecta unit in the Onverwacht Group of the BGB. The ~3.235 Ga unconformity between the volcanic Sulphur Springs Group, megabreccia and the clastic-felsic volcanic Soanesville Group in the central Pilbara Craton correlates within analytical error with the ~3.225 Ga unconformity between the Onverwacht Group and Fig Tree Group in the BGB. As in the BGB, the ~3.2 Ga break represents a fundamental change from the greenstone-TTG system to semi-continental environments dominated by arenites, turbidites, conglomerate, banded iron formations and felsic volcanics. These events resulted in a change of geotectonic patterns from pre-3.2 Ga largely dome-structured batholiths to linear structural patterns of supracrustal belts and intervening granitoids observed in the post-3.2 Ga western Pilbara Craton.

Keywords Asteroid impact • Western Australia • Crustal evolution • Pilbara Craton • Warrawoona Group • Sulphur Springs Group • Mega-breccia

8.1 Crustal Evolution

Reconnaissance geological mapping of parts of the ~60,000 km²-large northeast Pilbara Craton (Noldart and Wyatt 1962) and systematic 1:250,000 geological mapping (Lipple 1975; Hickman 1977, 1981, 1983, 1990; Griffin 1990; Trendall 1995) and geophysical survey the Geological Survey of Western Australia (GSWA) of the ~400,000 km²-large craton have established litho-stratigraphic subdivision

Fig. 8.1 Space shuttle image looking at the Pilbara region, northwestern Australia, from the north, showing the granite-greenstone patterns. NASA. *Inset* – Granite boulders, Yule Batholith

and correlations of the Archaean volcanic-sedimentary greenstone belts and major features of intervening diapiric granitic domes (Figs. 8.1 to 8.5). The preservation of a wealth of primary igneous and sedimentary structures and textures in least-deformed structural windows in the Pilbara Craton has allowed investigations of the original magmatic and sedimentary environments in which they have been deposited. Examples include pillow lavas, oscicular lavas (Fig. 8.6a), relic microscopic quench and porphyritic textures (Figs. 8.7[1], [2]), pyroclastic agglomerates (Fig. 8.8), variolitic spherulitic textures in felsic volcanics (Fig. 8.9), columnar lavas (Fig. 8.10a) and resorbed phenocrysts (Fig. 8.10b).

The geological mapping was accompanied by detailed structural, metamorphic, sediment logical, petrological, geochemical, isotopic and paleontological aspects by the University of Western Australia, Curtin University, Australian National University, Australian Geological Survey Organization (AGSO, now Geoscience Australia), interstate and overseas universities, in particular the University of Utrecht, Tokyo Institute of Technology, Stanford University and Oberlin College. Detailed 1:100,000 geological mapping since 1995 accompanied with airborne magnetic and radiometric surveys by GSWA greatly expanded field and laboratory research (Van Kranendonk et al. 2002, 2006, Van Kranendonk et al. 2007a, b; Hickman 2004; Hickman and Van Kranendonk 2004, 2008). Hickman (2012)

8.1 Crustal Evolution

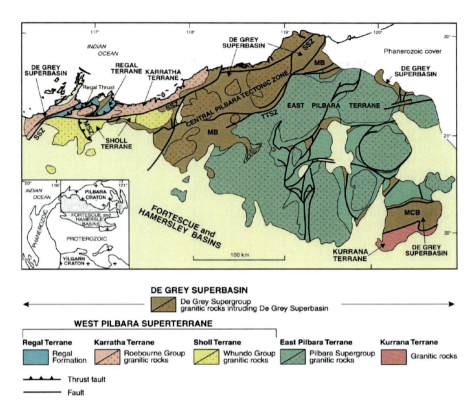

Fig. 8.2 Geological map of the northern Pilbara Craton showing terrains and the De Grey Superbasin. *MB* Mallina Basin, *MCB* Mosquito Creek Basin, *SSZ* Sholl Shear Zone, *MSZ* Maitland Shear Zone, *TTSZ* Tabba Tabba Shear Zone (Hickman 2012. Geological Survey of Western Australia, by permission)

reviewed the results of these studies (Hickman 1981; 2004, 2012; Hickman and Van Kranendonk 2004, 2008; Van Kranendonk 2010; Williams 1999, 2001; Williams and Bagas 2007; Smithies et al. 2005, 2007a, b; Trendall 1995; Glikson 1979a, b; Glikson and Hickman 1981 and other papers) in terms of the following evolutionary sequence (Appendix 2):

<3.7 Ga: Records include (1) 3.66–3.58 Ga biotite–tonalite enclaves within post-3.5 Ga granodiorite and monzogranite of Warrawagine gneisses, northeastern part of the Pilbara Craton (Williams 2001); (2) xenoliths of 3.58 Ga gabbroic anorthosite within ~3.43 Ga granitic rocks on the western margin of the Shaw Granitic Complex (McNaughton et al. 1988); (3) relic 3.72 Ga zircon within felsic units of the Pilbara Supergroup (Thorpe et al. 1992); (4) clastic 3.60 Ga zircons within Archaean sediments, including a 3.80 Ga zircon in the Mallina Basin and a 3.71 Ga zircon in the Mosquito Creek Basin; (5) Nd isotopic data from Archaean volcanics supporting relic pre-3.55 Ga ages (Jahn et al. 1981;

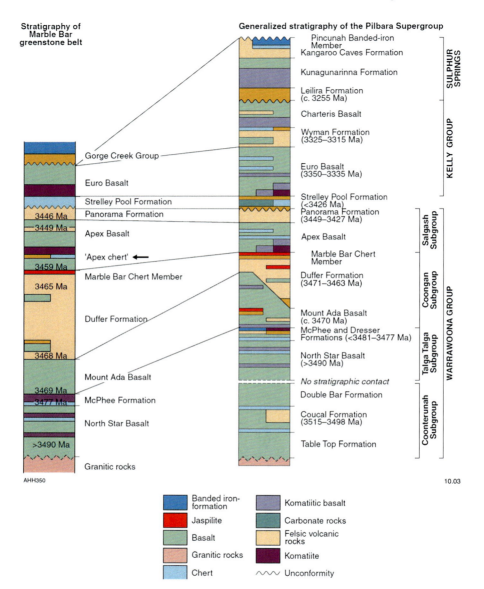

Fig. 8.3 Stratigraphy and isotopic ages of the Marble Bar greenstone belt. *Red dots* represent units correlated with stratigraphic units of the Barberton Greenstone Belt containing impact ejecta deposits. The *red dots* represent stratigraphic units correlated with ages of asteroid impact fallout units in the Barberton Greenstone Belt, South Africa (Geological Survey of Western Australia, by permission)

Van Kranendonk et al. 2007a, b; Tessalina et al. 2010); (6) evidence for formation of 3.50–3.42 Ga tonalite–trondhjemite–granodiorite (TTG) series from pre-3.5 Ga sources (Smithies et al. 2009).

Fig. 8.4 LANDSAT image of the eastern Pilbara Craton, showing the Mount Edgar (*ME*), Corruna Down (*CD*), Shaw (*S*) and Muccan (*MU*) batholiths and the North Pole dome (*NP*). The *inset* shows the North Pole Dome

3.53–3.235 Ga. Multiple successive volcanic cycles (Pilbara Supergroup) concomitant with plutonic granitic magmatism, culminating with formation of thick SIAL, are exposed in 18 greenstone belts in the eastern Pilbara Craton. The base of the Pilbara Supergroup is truncated by younger granites. The oldest exposed supracrustals represented by the >3.52 Ga Table Top Formation (Figs. 8.2 and 8.3) are intruded by 3.48–3.47 Ga granites of the Carlindi Granitic Complex (Buick et al. 1995). The pre-3.22 Ga terrain, estimated as 100,000 km^2-large, consists of a 15–20 km thick volcanic succession (Pilbara Supergroup) composed of at least eight ultramafic–mafic–felsic volcanic cycles, representing repeated mantle melting episodes. Repeated to semi-continuous partial melting events resulted in felsic magma accreted into growing batholiths. In some instances plutonic units are connected to felsic volcanic units by granitic veins, dykes, stocks, or cupolas, as in the Marble Bar greenstone belt where ~3,466 Ma granite intrudes an 8 km thick sequence of felsic volcanics of the 3,471–3,463 Ma Duffer Formation (Reynolds et al. 1975; Hickman and Lipple 1978; Hickman and Van Kranendonk 2008). The youngest pre-3.2 Ga-old volcanic cycle is represented by Sulphur Springs Group, which includes a major ultramafic-mafic volcanic unit–the Kunagunarinna Formation (~3,255–3,252 Ma) capped by felsic volcanics of the Kangaroo Cave Formation (3,253 ± 3) (Fig. 8.3).

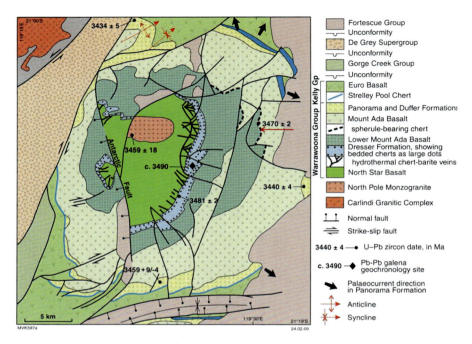

Fig. 8.5 Geological sketch map of the North Pole dome, central Pilbara Craton, Western Australia, showing the distribution of the ~3.47 Ga impactite unit, indicated by a *red arrow*, in the middle of the Mount Ada Basalt sequence in the northeastern part of the dome. Geological Survey of Western Australia, by permission

3.235 ± 3 Ga: An abrupt end of the volcanic mafic-ultramafic cycles is represented in the central Pilbara by the felsic volcanics of the Kangaroo Cave Formation of the volcanic Sulphur Springs Group (Van Kranendonk 2000), capped by a marker chert truncated by a major unconformity (Figs. 8.11, 8.16 and 8.17). Stratigraphic and geochronological correlations with the Barberton Greenstone Belt where an abrupt unconformable break occurs between the top of the volcanic Mendon Formation (3,258 ± 3 Ma) and the base of the clastic Mapepe Formation, accompanied with the S2 and S3 impact ejecta, suggest a connection with the effects of large asteroid impacts (Glikson and Vickers 2006) (Fig. 7.9). The unconformity is overlain by lenses of megabreccia (olistostrome) containing blocks up to 250 m-large (Fig. 8.16).

3.235 ± 3 Ga – 3.16 Ga: The Sulphur Springs Group is overlain by the clastic Soanesville Group (Corboy Formation, Paddy Market Formation, Pyramid Hills Formation) which includes volcanics of the Honeyeater Basalt, representing intracratonic sedimentation and volcanic activity (Figs. 8.2 and 8.3). At this stage rifting resulted in separation of the Karratha block (~3.50–3.25 Ga, Western Pilbara Craton) and Kurrana block (~3.50–3.17 Ga, southeast Pilbara Craton) from the Eastern Pilbara Craton. Evidence for development of SIMA crust between the East Pilbara terrain and the Karratha block is furnished by the ~3.2 Ga Regal Formation.

8.1 Crustal Evolution

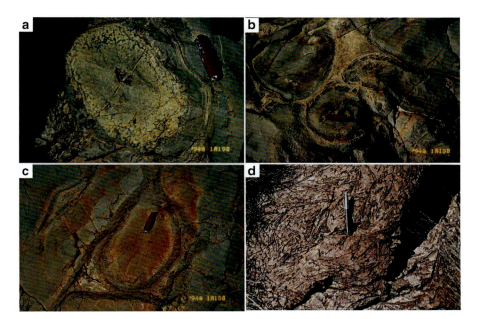

Fig. 8.6 Archaean volcanics, Pilbara Craton: (**a–c**) Pillow basalts, basal Apex Basalt, Coongan River, photo **a** showing outward boiling of vesicles and photos **b** and **c** showing inter-pillow fragmental mesostasis; (**b**) Quench "spinifex" textures in Archaean volcanics, Louden Basalt, west Pilbara Craton

3.13–3.11 Ga: Eruption of calc-alkaline volcanics (Whundo Group) and emplacement of granites in the western Pilbara Craton.

3.05–2.93 Ga: Arenite and granite pebble-bearing conglomerate deposited in the Gorge Creek, Whim Creek, Mallina, and Mosquito Creek Basins (Figs. 8.2 and 8.3) represent major uplift, exposure and denudation of Archaean batholiths throughout the eastern and western Pilbara Craton.

2.78–2.63 Ga: The Fortescue Basin, spanning 2.78–2.63 Ga, is represented by a ~6 km thick sequence of volcanic flows dominated by basaltic andesite, with intercalated carbonate (Tumbiana Formation), and terminated by extensive deposition of shale, chert and BIF of the 2.69–2.63 Ga Jeerinah Formation (Thorne and Trendall 2001). The Fortescue Group unconformably overlies the Pilbara Craton over an area of at least 250,000 km² (Blake 1993). Three mafic-felsic volcanic cycles are observed (Mount Roe Basalt, Hardey Formation, and Kylena to Maddina Formations) (Kojan and Hickman 1998), interpreted in terms of mantle plume events by Arndt et al. (2001). The Mount Roe Basalt, reaching thicknesses of up to 2,500 m (Blake 1993), is preserved mainly in palaeo-valleys which overlie subsiding Archaean greenstone belts subsiding between batholiths due to differential isostatic movements. The feeders are represented by north-northeast trending dolerite dykes (Black Range Dolerite Suite). In the northern Pilbara basaltic volcanism was succeeded with formation of shallow rift valleys

104　　　　　　　　　　　　　　　　　8　Evolution and Pre-3.2 Ga Asteroid Impact Clusters…

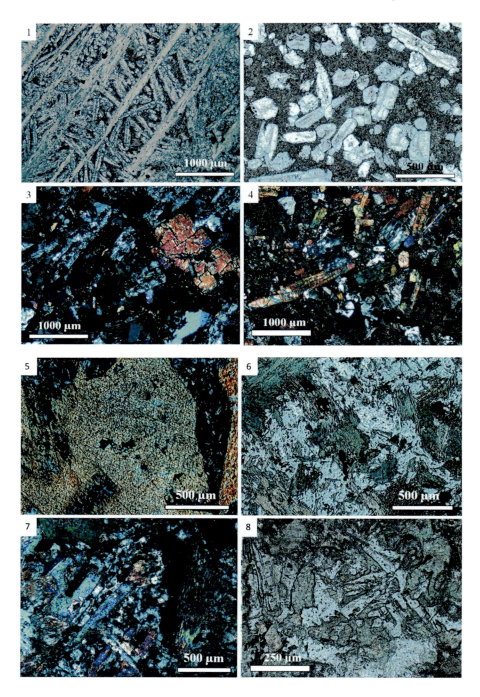

8.1 Crustal Evolution

Fig. 8.8 Jaspillite fragment-bearing volcanic agglomerate, Duffer Formation, Coongan River, Marble Bar

filled with clastic sediments of the 2.76–2.75 Ga Hardey Formation (Thorne and Trendall 2001). The second volcanic cycle is represented by high-K basalts of the 2.74 Ga Kylena Formation, overlain by felsic tuff and stromatolitic carbonate of the lacustrine 2.73–2.72 Ga Tumbiana Formation, and a third volcanic cycle by the 2.71 Ga Maddina Formation (Sakurai et al. 2005; Hickman 2012). The Fortescue Group terminates at 2.69 Ga with extensive deposition of off-shore mudstone, pyritic shale, siltstone, chert, carbonates of the Jeerinah Formation, representing a northerly marine transgression in a deepening basin (Thorne and Trendall 2001).

Hickman (2012) classified the northern Pilbara Craton in terms of five 3.53–3.11 Ga-old granite–greenstone terrains and a number of volcanic and sedimentary

Fig. 8.7 Microscopic images of Archaean volcanics, Apex Basalt, Camel Creek, East Pilbara Craton: (*1*) spinifex-textured komatiite consisting of skeletal pseudomorphs of tremolite after pyroxene and interstitial altered feldspathic matrix; (*2*) porphyry, displaying crystals of altered plagioclase set in cryptocrystalline chlorite-rich matrix; (*3*) orthopyroxene microphenocrysts set in clouded ophitic-textured altered basalt; (*4*) Clinopyroxene prisms set in clouded basaltic matrix; (*5*) gabbro – a phenocryst of clinopyroxene enclosing a plagioclase in ophitic relations; (*6*) metamorphosed gabbro displaying acicular actinolite set in albite-oligoclase; (*7*) dolerite – lath-like crystals of plagioclase separated by clouded microcrystalline clinopyroxene-bearing matrix; (*8*) ophitic to graphic-textured actinolite-plagioclase dolerite. (Figures *3, 4, 5* and *7* are in cross nicols)

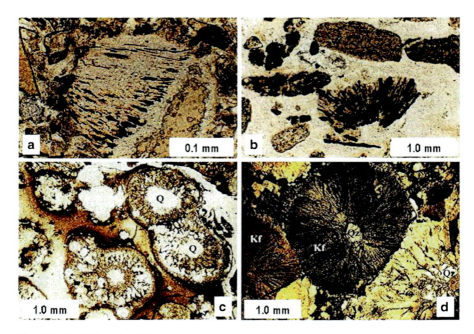

Fig. 8.9 Variolitic spherulitic textures in Archaean felsic volcanics, lower part of Soanesville Group, East Pilbara Craton. Note the distinction between the outward-radiating crystallites of the volcanic spherules, in some instances cored by feldspar crystal, and microkrystite spherules displaying inward-radiating crystallites (Figs. 7.12, 7.13, 10.3e, 10.4 and 10.7d) (From Glikson and Vickers 2006; Elsevier, by permission)

basins. From ~3.22 Ga onwards, these terrains and basins evolved separately in different parts of the Pilbara Craton (Figs. 8.2 and 8.3). These terrains and basins include (Hickman 2012):

1. 3.22–3.17 Ga Soanesville Basin and similar rift-related basins occupied by epicontinental clastic successions overlain by deeper water volcanics along the rifted margins of the East Pilbara, Karratha, and Kurrana Terrains.
2. ~3.2 Ga Regal Terrain, West Pilbara, represents mafic crust formed between the East Pilbara and Karratha plates later than ~3.22 Ga.
3. 3.13–3.11 Ga Sholl Terrain, consisting of ~10 km-thick volcanics of intermediate composition (Whundo Group) and related granitic rocks.
4. 3.05–2.93 Ga De Grey Superbasin, extensional basins intruded by granitic plutons between 3.02 and 2.94 Ga.
5. 2.89–2.83 Ga Split Rock Supersuite, post-tectonic granitic intrusions

Prior to 3.22 Ga the three major greenstone successions (Warrawoona, Kelly, and Sulphur Springs), dominated by mafic-ultramafic volcanic sequences capped by felsic volcanic sequences (Duffer, Panorama, Wyman, and Kangaroo Caves formations), are separated by major unconformities (Figs. 8.2 to 8.5). The volcanic sequences and contemporaneous granitic supersuites (Callina, Tambina, Emu Pool, and Cleland)

8.1 Crustal Evolution

Fig. 8.10 (**a**) columnar rhyolite of the Wyman Formation (3,325–3,215 Ma); (**b**) resorbed quartz crystal in the Wyman Formation rhyolite, signifying derivation from depths where quartz phenocrysts are stable in melts, as discussed in Green and Ringwood (1977)

represent coeval, in some instances connected, extrusive and plutonic anatectic events associated with or succeeding mantle melting events. The significant time intervals separating the Warrawoona and Kelly Groups (3,427–3,350 Ma; ~75 Ma) and the Kelly and Sulphur Springs Groups (3,315–3,255 Ma; 60 my) may represent deformation and metamorphism of underlying sequences, long-term hiatuses and missing sequences. The Warrawoona and Kelly groups are separated by an unconformity overlain by ~1,000 m-thick shallow water sediments of the Strelley Pool Formation (Hickman and Van Kranendonk 2008) (Fig. 8.3). The Kelly and Sulphur Springs groups are separated by unconformity overlain by up to 3,900 m-thick clastic sediments of the Leilira Formation (Van Kranendonk and Morant 1998; Buick et al. 1995, 2002; Hickman 2012).

Fig. 8.11 Pilbara Craton, >3.43 Ga and ~3.2 Ga unconformities. (**a**) northern limb of the Strelley anticline enveloping the Strelley Granite (*SG*). *Arrows*: 1 – referring to inset *B*; 2 – referring to inset *C*; 3 – unconformity at the base of ~3.0 Ga Gorge Creek quartzite and conglomerate; (**b**) unconformity between the Coonterunah Group (3,515–3,498 Ma) and the overlying Strelley Formation (~3,426 Ma); (**c**) basal Soanesville Group olistostrome unconformably overlying Sulphur Springs Group and overlain by turbidite-felsic volcanic sequence. For stratigraphic relations and ages see Fig. 8.3

The youngest greenstone cycle of the 3.53–3.23 Ga Pilbara Supergroup, the Sulphur Springs Group, is terminated abruptly at ~3.23 Ga, where the uppermost felsic volcanic unit, Kangaroo Cave Formation (~3,252–3,235 Ma) is truncated unconformably by the Soanesville Group (Fig. 8.16) which is dominated by clastic sediments, banded iron formation, felsic volcanics and conglomerate accompanied by felsic tuff and basalt (Hickman 2012), signifying the onset of semi-continental conditions. The base of the Soanesville Group includes lenses of mega-breccia (olistostrome) consisting of blocks up to 250 m-large (Hill 1997) (Fig. 8.16). Equivalents of the Soanesville Group and Nickol River Formation are represented in the southeast Pilbara by sediments of the ~3.22 Ga Budjan Creek Formation (Fig. 8.12) and associated volcanics of the Kelly Belt, representing intra-cratonic rifting (Erikson 1981). Equivalents of the Soanesville Group also occur in the Kurrana terrain (Fig. 8.2) where the >3.18 Ga Coondamar Formation meta-sediments and volcanics are interleaved with granites. In the western Pilbara correlated

Fig. 8.12 Pilbara Craton ~3.3–3.2 Ga unconformities, Budjan Creek. Aerial photograph displaying high-angle unconformity between the Wyman Rhyolite (*WR* – 3,308 ± 4 Ma) and the Budjan Creek Formation (BCF – 3,228 ± 6 Ma), Kelly greenstone belt. *UNC* unconformity (NASA)

formations include the ~3.3–3.27 Ga Ruth Well basalts and unconformably overlying clastics, quartzite, chert shale, BIF, and carbonates of the ~3.22 Ga Nickol River Formation (NRF). The NRF is overlain by the ~3.2 Ga (Van Kranendonk 2007a, b; Kiyokawa et al. 2002), 2–3 km-thick volcanic Regal Formation, consisting of pillow basalt, basal komatiitic peridotite and rare chert, with possible geochemical signatures of oceanic crust (Ohta et al. 1996; Sun and Hickman 1998; Smithies et al. 2005, 2007a, b). During about 3.16–3.07 Ga the Central Pilbara Craton separated from the Karratha terrain of the west Pilbara and the Kurrana Terrain of the southeast Pilbara, underlain by SIMA crust represented by Regal Formation. The development of these northeast-trending Terrains (Van Kranendonk et al. 2006; Eriksson 1981; Wilhelmj and Dunlop 1984) was accompanied with intrusion of ~3.07 Ga granites, signifying ~3.18 Ga rifting subsidence of thick clastic sequences accompanied by intrusion of mafic dykes and sills of dolerite and gabbro, as well as intrusion of 3.19–3.17 Ga granites of the Kurrana Terrain. Interpretations by workers of the Geological Survey of Western Australia regard the ~3.5–3.2 Ga greenstones as a volcanic plateau formed above continental basement, followed at !3.2 Ga by development of passive margin and non-continental basins prior to 3.07 Ga.

The younger volcanic terrains in the western Pilbara Craton are represented by >6 km-thick volcanic sequences of the 3.13–3.11 Ga Whundo Group and the 3.13–3.11 Ga

Railway Supersuite of the Sholl Terrain, consisting of meta-basalt, ultramafic rocks, intermediate pyroclastic rocks, andesite, dacite, rhyolite, dolerite sills and thin metasedimentary units, (Hickman 2012). The Whundo Group includes calc-alkaline and boninite-like rocks, a middle package of tholeiitic rocks with minor boninite-like rocks and rhyolites, and an upper package of calc-alkaline rocks, including adakites, Mg-rich basalts, Nb-enriched basalts, and rhyolites, with Nd isotopic compositions close to depleted mantle values (εNd of +1.5–2.0 at ~3,120 Ma) (Sun and Hickman 1998; Smithies et al. 2005). On this basis several authors compared the Whundoo Group to modern convergent margins, arc-trench, back-arc systems or intra-oceanic systems (Krapez and Eisenlohr 1998). The low Th/La, La/Nb, and Ce/Yb ratios of boninites are consistent with intra-oceanic arc setting (Smithies et al. 2005). Associated plutonic units, including meta-tonalite, Meta-granodiorite and meta-monzogranite gneiss, contain zircons up to 3,149 Ma-old and Nd TDM model ages of 3.23–3.21 Ga (Smithies et al. 2007a, b).

Uplift of pre-3.1 Ga SIAL blocks was followed by an opening of oceanic gaps at ~3.13–3.11 Ga, represented by the volcanic Whundo Group, tectonic movements (~3.07 Ga "Princep Orogeny") and development of major up to 2,000 m thick shallow-water molasse-type conglomerates, fanglomerates and quartzite (De Grey Superbasin) in the northern Pilbara Craton, unconformably overlying the ~3.6–3.2 Ga granite greenstone system (Dawes et al. 1995). Post 3.07 Ga continental basins include the 3.05–3.02 Ga Gorge Creek Basin with massive conglomerates of the Lalla Rookh Sandstone (Eriksson et al. 1994), the 3.01–2.99 Ga Whim Creek Basin, the 3.01–2.94 Ga Mallina Basin and the 2.97–2.93 Ga Mosquito Creek Basin (Hickman et al. 2010; Hickman 2012). The correlated 3.01–2.99 Ga Whim Creek Basin in the west, containing felsic and basaltic volcanics, was regarded as an ensialic back-arc basin (Pike and Cas 2002). Plutonic equivalents of the Whim Creek Group are likely represented by the 3.00–2.98 Ga Maitland River Supersuite of granites. The Mallina Basin contains a 2–4 km-thick sequence of conglomerate, arenite, shale and intercalated basalts and high-Mg basalt (Eriksson 1982; Smithies 2004). Plutonic and volcanic activity during 2.95–2.94 Ga (Negri and Louden volcanics) is interpreted in terms of subduction (Smithies and Champion 2000). The Mosquito Creek Basin in the southeast Pilbara Craton (2.95–2.94 Ga), formed in subsiding rift structures, contains graded-bedded arenite and siltstones displaying Bouma cycle sequences (Van Kranenkonk et al. 2006). Post-orogenic plutonic activity is represented by 2.89–2.83 Ga fractionated monzogranite and syenogranite (Split Rock Supersuite).

8.2 ~Pre-3.2 Ga Impact Fallout Units in the Pilbara Craton

A geological map of the Pilbara Craton indicating the distribution of impact ejecta units is presented as a pdf enclosed with the book, a condensed version of which is shown in Fig. 8.13 (http://extras.springer.com/2014/978-3-319-07908-0).

Lowe and Byerly (1986b) reported millimeter to sub-millimeter scale silicified spherules containing quench-crystallization and glass-devitrification textures from a

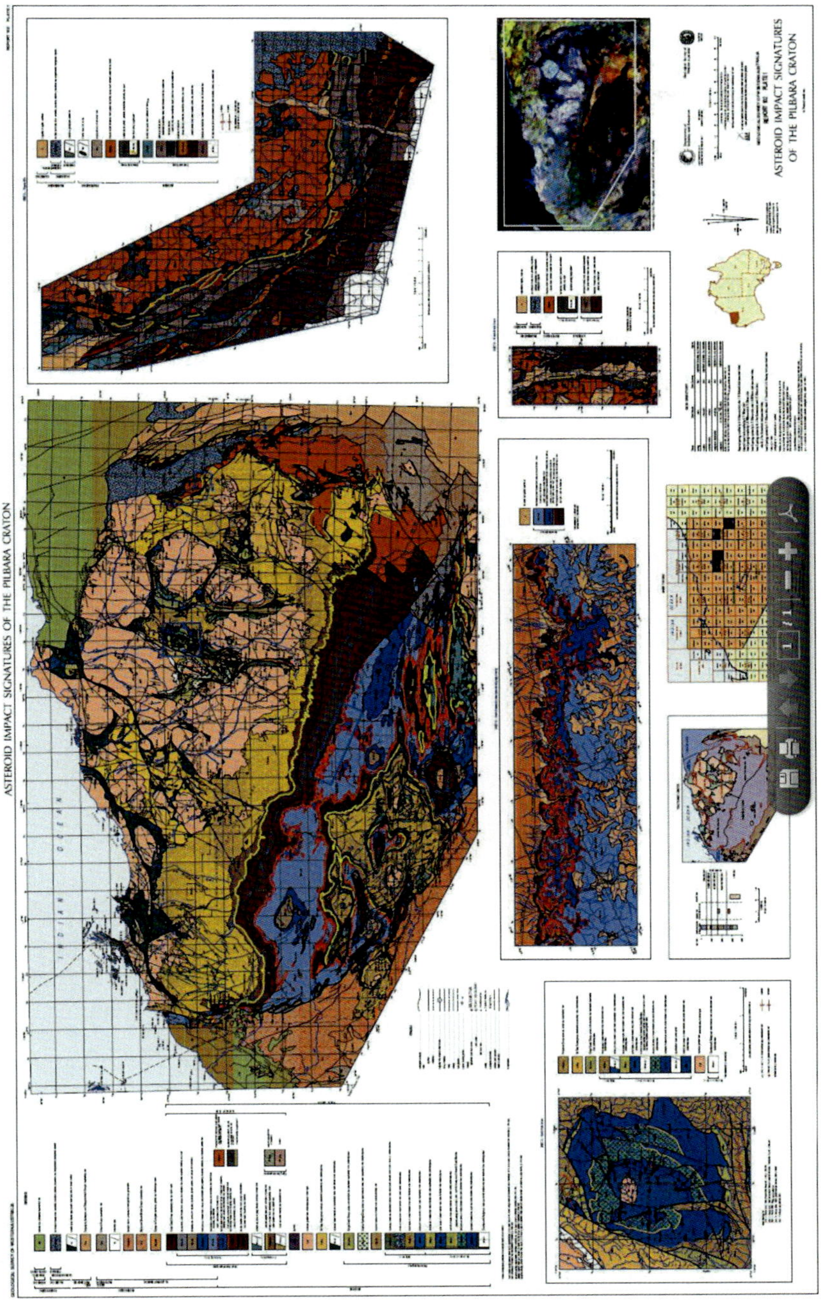

8.2 ~Pre-3.2 Ga Impact Fallout Units in the Pilbara Craton 111

Fig. 8.13 Pilbara asteroid impact map (Glikson 2004a, b. Geological Survey of Western Australia, by permission. pdf file on line http://extras.springer.com/2014/978-3-319-07908-0)

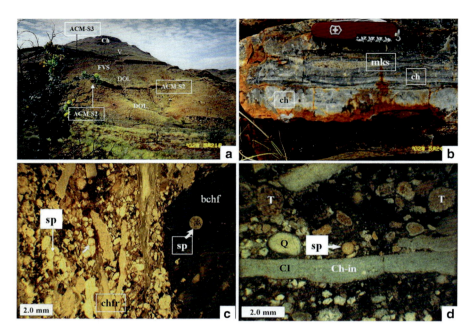

Fig. 8.14 Pilbara Craton ~3.47 Ga impact spherules. Antarctic Chert Member impact unit. (**a**) View of type locality north of Miralga Creek. Abbreviations: *DOL* dolerite, *ACM-S2* spherule-bearing diamictite, *FVS* felsic volcanics/hypabyssals, *ACM-S3* spherule-bearing chert and arenite, *FV* felsic volcanics, *Ch* chert. (**b**) ACM-S3 – spherule lens (*mks*) in laminated chert (*ch*). Swiss knife – 8 cm; (**c**) ACM-S3 – chert intraclast diamictite, including bedding-parallel chert fragments (*chfr*) and a black chert fragment (*bchf*) which includes a microkrystite spherule (*sp*) – possibly derived from an older spherule unit ACMS1; (**d**) ACM-S3 chert-intraclast diamictite including bedding relic chert intraclast (*Ch-in*) and spherules in the matrix (*sp*)

chert/arenite unit located about 3.0 km above the Dresser Formation, tracing this unit about 1 km along strike (Figs. 8.14 and 8.15). The underlying Mount Ada Basalt, as defined by Hickman (1983, 2012) and Van Kranendonk (2000) Van Kranendonk and Morant (1998), consists of extensively carbonated pillowed tholeiitic to high-Mg basalt, dolerite sills, minor volcaniclastic rocks and ~1–10 m-scale intercalations of chert, chert/arenite and minor intraclast chert pebble-dominated conglomerate. In the North Shaw 1:100,000 Sheet area, the Mount Ada Basalt forms a ~6.5 km-thick sequence on the eastern part of the North Pole dome (Figs. 8.4 and 8.5), overlying the Dresser Formation (chert, arenite, barite) and underlying the Panorama Formation (carbonate-altered felsic volcanic and volcaniclastic rocks and derived clastic sedimentary rocks). The isotopic age of the Mount Ada Basalt is constrained in the adjacent Marble Bar greenstone belt to the east by the age of the underlying Duffer Formation (~3,471–3,463 Ma) and the age of the overlying Panorama Formation (~3,458–3,454 Ma) (Van Kranendonk et al. 2002; Hickman 2012). The thickest (5–20 m) sedimentary intercalation within the Mount Ada Basalt, titled Antarctic Creek Member, consists of felsic volcani-clastics, mafic tuff silicified argillite, aren-

8.2 ~Pre-3.2 Ga Impact Fallout Units in the Pilbara Craton

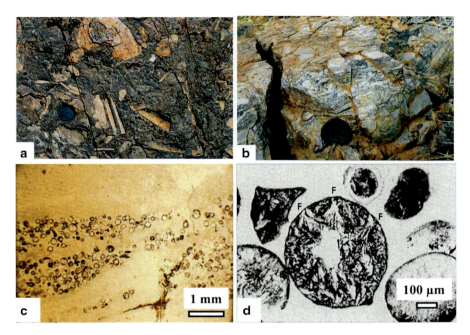

Fig. 8.15 Pilbara Craton ~3.47 Ga impact spherules. ~3.47 Ga Antarctic Chert Member impact unit (**a**) ACM-S2 – chert figment diamictite, viewed parallel to bedding plain, showing some pebbles at high angle to the layering. Camera lens cap (*LC white arrow*) – 6 cm; (**b**) chert fragment diamictite viewed across the layering; (**c**) a pocket of silicified microkrystite spherules; (**d**) silicified microkrystite spherule preserving relic inward-radiating pseudo morphed crystallites

ite, Jaspilite intruded by dolerite (Fig. 8.14). Above this unit, the volcanic sequence includes several intercalations of chert/arenite and intraclast flat chert-pebble conglomerate. A similar spherules-bearing fragmental unit near 40 meters-thick has recently bee identified under the Marble Bar Chert about 25 km to the east of the Miralga Creek occurrence.

Spherules are mostly 0.10–0.75 mm in diameter. Irregular particles, compound spherules, dumbbell shaped particles and broken spherules are present. Spherules within chert retain high to very high sphericity and internal radiating quench textures, allowing their positive identification as microkrystite spherules (Glikson et al. 2004) as defined by Glass and Burns (1988). Byerly et al. (2002) reported a 207/206 Pb age of 3,470.1 ± 1.9 Ma for euhedral zircons derived from the spherule-bearing unit, i.e. within error from the age of the Duffer Formation (3,467 ± 4, Van Kranendonk et al. 2002). Of the 30 zircons of 50–100 μm extracted from 2 kg of rock, two grains yielded ages of c. 3,510 Ma, suggesting derivation of some of the clastic material from the older Coonterunah Group (Buick et al. 1995; Van Kranendonk and Morant 1998; Van Kranendonk et al. 2002). Byerly et al. (2002) correlate the spherule-bearing unit with a similar S1 unit in the Hoogenoeg Formation, Barberton Greenstone belt, Kaapvaal Craton, South Africa, which yielded a 207/206 Pb zircon age of 3,470.4 ± 2.3 Ma (Sect. 7.3).

The microkrystite-bearing units of the Antarctic Chert Member, defined here as ACM-S, consist of at least three horizons, ACM-S1-3, including (Glikson et al. 2004):

1. ACM-S1 – Spherule-bearing chert fragments in ACM-S3.
2. ACM-S2 – A basal lens of spherule-bearing diamictite, located below ACM-S3 from which it is separated by a ~200 m-thick dolerite and ~30 m-thick felsic hypabyssals.
3. ACM-S3 – A chert unit bearing spherules and chert fragments;

The microkrystite spherules are discriminated from angular to subangular detrital volcanic fragments by their high sphericities, inward-radiating fans of sericite pseudomorphs after K-feldspar, relic quench textures and Ni-Cr-Co relations. Scanning Electron Microscopy coupled with E-probe (EDS) and laser ICPMS analysis indicate high Ni and Cr in sericite-dominated spherules, suggesting mafic composition of source crust. Ni/Cr and Ni/Co ratios of the spherules are higher than in associated Archaean tholeiitic basalts and high-Mg basalts, rendering possible contamination by high Ni/Cr and Ni/Co chondritic components. The presence of multiple bands and lenses of spherules within chert and scattered spherules in arenite bands within S3 may signify redeposition of a single impact fallout unit or, alternatively, multiple impacts. Controlling parameters include: (1) spherule atmospheric residence time; (2) precipitation rates of colloidal silica; (3) solidification rates of colloidal silica; (4) arenite and spherule redeposition rates, and (5) arrival of the tsunami. The presence of spherule-bearing chert fragments in S3 may hint at an older spherule-bearing chert (ACM-S1). Only a minor proportion of spherules are broken and the near-perfect sphericities of chert-hosted spherules and arenite-hosted spherules constrain the extent of shallow water winnowing of the originally delicate glass spherules. It follows the spherules were either protected by rapid burial or, alternatively, disturbance was limited to a short term high energy perturbation such as may have been affected by a deep-amplitude impact-triggered tsunami wave.

8.3 ~3.24–3.227 Ga Impact-Correlated Units

Three impact spherule-bearing units (S1–3) were identified on the basis of quench textures, iridium anomalies and Cr isotopic data in the Barberton Greenstone Belt (BGB) at the base to lower part of the Fig Tree Group (Lowe and Byerly 1986a, b; Lowe et al. 1989, 2003; Byerly and Lowe 1994; Byerly et al. 2002; Kyte 2002; Kyte et al. 1992, 2003; Shukolyukov et al. 2000; Glikson 2007). Significant stratigraphic, lithological and isotopic age analogies, as well as differences, pertain between sequences in the BGB and Eastern Pilbara greenstone belts, including likely correlations between the following stratigraphic/unconformity breaks about ~3.25–3.22 Ga (Glikson and Vickers 2006, 2010) (Figs. 8.3 and 8.17)

(A) In the BGB the hiatus between the upper part of the ultramafic to mafic/felsic volcanic Onverwacht Group (Mendon Formation 3,298 ± 3 Ma) and turbidite/felsic volcanics of the lower Fig Tree Group (Mapepe Formation – 3,258 ± 3–3,225 ± 3 Ma) in the Barberton Greenstone Belt (U-Pb zircon ages after Kroner et al. 1991a, b and Byerly et al. 1996).

8.3 ~3.24–3.227 Ga Impact-Correlated Units   115

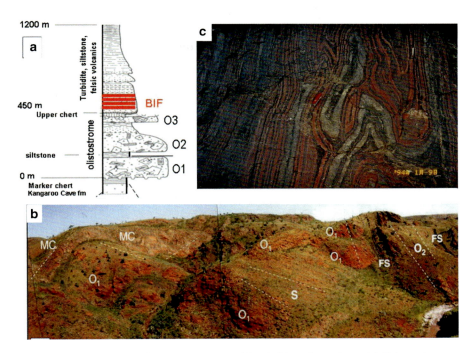

Fig. 8.16 Pilbara Craton ~3.2 olistostrome and banded iron formations: (**a**) Schematic cross section showing the basal unconformity of the olistostrome over volcanics of the Sulphur Springs Group, multiple olistostrome units and overlying banded iron formations, turbidites and felsic volcanics of the Soanesville Group; (**b**) A view of olistostrome units (O₁, O₂, O₃), intercalated siltstone (*S*) and ferruginous siltstone (*FS*) and the underlying Marker Chert (*MC*) signifying the top of the Sulphur Springs Group; (**c**) Banded iron formation of the Gorge Creek Group (From Glikson and Vickers 2006; Elsevier by permission)

(B) In the central Pilbara Craton a break occurs between the upper part of the ultramafic to felsic volcanic Sulphur Springs Group (U-Pb zircon ages of 3,255–3,235 Ma, Van Kranendonk and Morant 1998; Van Kranendonk 2000; Buick et al. 2002) and overlying mega-breccia (olistostrome), ferruginous argillite/turbidite/felsic volcanic BIF (banded iron formation)-bearing Gorge Creek Group, Pilbara Craton (Figs. 8.16 and 8.17).

The proposed correlation between these breaks has led to a search for impact fallout units in corresponding units in the Pilbara Craton (Glikson and Vickers 2006) (Fig. 8.17). The close stratigraphic and geochronological correlations between (1) the ~3.258 Ga break between the Onverwacht Group–Mendon Formation and Fig Tree Group, and (2) breaks at ~3.255 and 3.235 Ga in the central Pilbara Craton on the other hand, support an investigation of the latter in terms of the effect of large asteroid impacts. The ~3.255–3.235 Ga Sulphur Springs Group (SSG) overlies the ~3.35–3.31 Ga Kelly Group unconformably and is dominated by a volcanic succession that varies in composition upward from a basal unit of conglomerate, wackes and

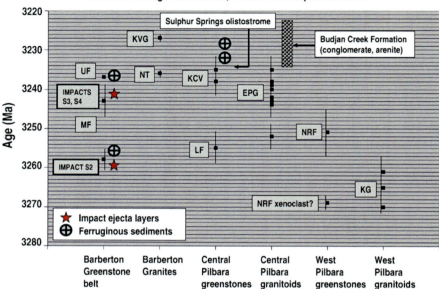

Fig. 8.17 Isotopic age correlations between 3.28 and 3.22 Ga units in the Kaapvaal Craton and Pilbara Craton. *Solid squares* and *error bar lines* – U–Pb ages of volcanic and plutonic units; *stars* – impact ejecta layers; *circled crosses* – ferruginous sediments; *MF* Mapepe Formation, *UF* Ulundi Formation, *NT* Nelshoogte tonalite, *KVG* Kaap Valley Granite, *LF* Leilira Formation, *KCV* Kangaroo Cave volcanics, *EPG* Eastern Pilbara granites, *NRF* Nickol River Formation (including an older xenoclast), *KG* Karratha Granite (From Glikson 2008; Elsevier, by permission)

felsic volcanic rocks (~3,255 Ma Leilira Formation), through komatiite (Kunagunarinna Formation), to a unit of basalt-andesite and rhyolite, with interbedded chert horizons (3.235 Ga Kangaroo Caves Formation). The volcanic rocks are overlain by a chert horizon, up to 50 m thick, composed of silicified fine-grained epiclastic and siliciclastic rocks. The chert formed during hydrothermal circulation associated with the final emplacement of the syn-volcanic Strelley Granite laccolith and precipitation of volcanogenic massive sulphide deposits (Vearncombe et al. 1998; Brauhart et al. 1998). The chert is overlain in turn by an olistostrome breccia (Fig. 8.16) and by fine-grained clastic and locally chemical sedimentary rocks of the Pincunah Hill Formation (Gorge Creek Group), including felsic tuff lenses. Marked sedimentary facies variations are observed in these sediments, which include (Van Kranendonk and Morant 1998; Van Kranendonk 2000). These units include:

1. A polymictic megabreccia, or olistostrome, reaching up to 600 m thick across a 4 km long strike length and consisting of blocks of chert, ferruginous shale, argillite, banded iron formation and felsic volcanic rocks set in an arenitic to ruditic matrix dominated by pyroclastic felsic volcanic material, mapped in detail by Hill (1997) and described by Vearncombe et al. (1998) and Van

Kranendonk (2000), unconformably overlies a marker chert at the top of the Sulphur Springs Group. The unusual dimensions of the blocks, up to <250 m, and their angular geometry militate for major faulting.
2. Ferruginous argillite, shale, and banded iron formation, including thin lenses of felsic pyroclastic rocks (Pincunah Hill Member). The Pincunah Hill Member varies laterally in thickness and pinches out altogether at the apex of the synvolcanic Strelley Granite. At this locality, the overlying Corboy Formation fills in a submarine canyon as a series of turbidites that lap onto the marker chert which had previously been lithified, as evidenced by Neptunian sandstone dykes filled by Corboy sandstones in the marker chert (Van Kranendonk 2000).
3. Turbiditic feldspathic arenite beds grading up to argillite (Corboy Formation). Thickness variations in the Corboy Formation to the west of the Sulphur Springs area occur across faults, suggesting horst and graben fault activity during sedimentation (Wilhelmij and Dunlop 1984).

A search for impact fallout unit/s in sedimentary intercalations of 3.255–3.235 Ga exposed units and drill cores has not to date identified microkrystite-bearing impact fallout units. Volcanic varioles and amygdales are widespread in felsic to intermediate metavolcanic rocks and in pyroclastic rocks and are distinguished from impact spherules by (1) occurrence of resorbed quartz microphenocrysts in the volcanic fragments (Fig. 8.9); (2) outward-radiating textures in varioles, as contrasted with inward radiating and quench textures in impact-condensation spherules (microkrystites) (Figs. 7.12, 7.13, 10.3e and 10.7); (3) poor sorting and occurrence of >5 mm-scale spherules in volcanic deposits, contrasted with the highly uniform size distribution and mostly mm to sub-mm-scale of impact spherules.

It is likely that the arenite-dominated composition of the Leilira Formation and part of the basal Soanesville Group resulted in corrosion and destruction of the originally glassy microkrystite spherules. The lenticular geometry of impact fallout units, probably breflecting syn-depositional current-induced erosion/redistribution of the spherules, requires further search for impact fallout lenses in sedimentary intercalations of the Sulphur Springs Group, including (1) the chert at the top of the Sulphur Springs Group; (2) the olistostrome breccia unit above the marker chert unit at Sulphur Springs; (3) Pincunah Hills Formation, and (4) base and lower parts of the arenitic Corboy Formation. The olistostrome may have been associated with the end of felsic volcanic activity and caldera collapse (Van Kranendonk et al. 2002); however, the location of the megabreccia/olistostrome at a stratigraphic level correlated with the Barberton impact spherules, and the several hundred meters-size of the blocks (olistoliths), conceivably hint at contemporaneous seismic/earthquakes effects that may relate to impact energy. Further study of the matrices of the olistostrome and intercalated argillite units is planned.

Chapter 9
Post-3.2 Ga Granite-Greenstone Systems

Abstract By analogy to pre-3.2 Ga system the younger granite-greenstone terrains display overall trends of evolution from largely mafic and ultramafic volcanic sequences to clastic sedimentary sequences, including turbidites and conglomerates. By contrast to pre-3.2 Ga granite-greenstone systems, commonly of oval or domal structure, post-3.2 Ga Archaean systems are mostly linear to sub linear, likely reflecting lateral accretion processes, for example supported by progressive southward isotopic age zonation of greenstone belts in the Superior Province. The linear structural grain of post-3.2 Ga systems is shown by seismic reflection studies in the Yilgarn Craton to be related to thrust and low angle detachment faults. However, a strict comparison between upper to late Archaean greenstone belts and circum Pacific-like accretionary wedges is negated by the lack of ophiolite-melange wedges of the type described along the northeastern Pacific rim (Hamilton, GSA Today 13:412, 2003). As is the case for pre-3.2 Ga systems, post-3.2 Ga mafic-ultramafic volcanics may represent rifted oceanic-like crustal zones developed along or between older gneiss terrains. In the Yilgarn Craton, southwestern Western Australia, such older gneiss terrains are represented by the Narryer terrain, southwestern gneiss terrain and by detrital zircons indicating pre-existing gneisses, the latter including peak magmatism about 3.2–3.3 Ga, a period correlating with the Barberton asteroid impact cluster.

Keywords Yilgarn Craton • Dharwar Craton • Superior Province • Linear greenstone belts • Lateral accretion

9.1 Evolution of the Yilgarn Craton

The Archaean Yilgarn Craton, southwestern Australia (Figs. 9.1 to 9.4), consists of metavolcanic and meta-sedimentary belts, granites and gneiss (Kinny et al. 1990; Pidgeon and Wilde 1990; Nelson 1997; Pidgeon and Hallberg 2000; Wyche 2007;

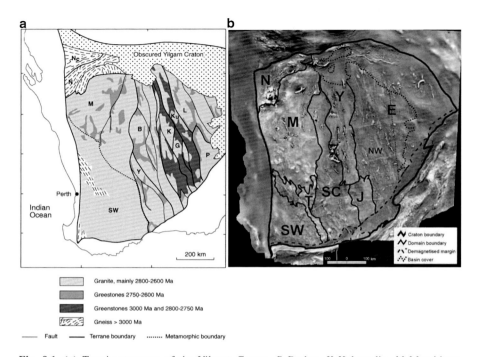

Fig. 9.1 (a) Terrain structure of the Yilgarn Craton. *B* Barlee, *K* Kalgoorlie, *M* Murchison, *N* Narryer, *Nc* Narryer terrain affected by the Capricorn Orogeny, *SW* southwest Yilgarn composite terrain, *Y* Yellowdine, *Ku* Kurnalpi, *G* Gindalbie, *L* Laverton, *P* Pinjin (From Griffin et al. 2004; Elsevier by permission); (b) Total magnetic intensity image of the Yilgarn Craton – *black* to *white* represents low to high magnetization. Geophysical domains are labelled *N* Narryer, *M*– Murchison, *T* Toodyay-Lake Grace, *SW* Southwest, *SC* Southern Cross, *Y* Yeelirrie, *J* Lake Johnston, *E* Eastern Goldfields. NW in the west of the Eastern Goldfields domain correlates approximately with the Norseman-Wiluna Belt of Gee et al. (1981) (Whittaker 2001)

Van Kranendonk et al. 2013) which represent an assembly of separate blocks amalgamated into the craton during 2.76–2.62 Ga, a period of extensive granitic plutonic activity and east-west compression (Nutman et al. 1997; Myers 1997; Myers and Swager 1997). The craton constitutes the most extensive remnant of Archaean continental crust in Australia, displaying a long history of crustal fragmentation and aggregation by continental collision and accretion (Myers 1990), including the collision of the Pilbara and Yilgarn cratons to form the intervening Capricorn orogen between 2.000 and 1.600 Ga and thrusting along the southeastern margin of the Yilgarn Craton to form the Albany-Fraser orogen between 1.300 and 1.100 Ga. Rifting removed the northeastern part of the Yilgarn and Pilbara cratons between 1.1 and 0.7 Ga and further rifting resulted in the Paterson orogen about 0.7–0.6 Ga. The oldest recorded components include 3.73 Ga layered megacrystic anorthosites and associated amphibolite, intruded by sheets of ~3.6 Ga granite in the northwestern Narryer Terrain (Myers 1997). The mafic bodies were associated with 3.5–3.3 episodes of high grade metamorphism, deformation and melting. Central provinces of

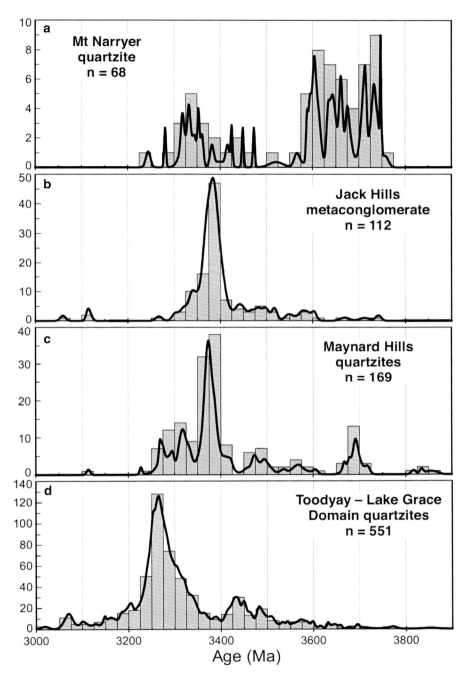

Fig. 9.2 Distribution of detrital zircon ages in the Yilgarn Craton between 3.0 and 3.9 Ga. (**a**) Quartzite from Mt Narryer in the Narryer Terrain (*NT*), (**b**) metaconglomerate from the Jack Hills greenstone belt in the *NT*, (**c**) quartzites from the Maynard Hills greenstone belt, Southern Cross Domain (*SCD*), and (**d**) quartzites from the Toodyay-Lake Grace Domain (*TLGD*) of the South West Terrain. All data are <10 % discordant; ages (from duplicate analyses of the same zircons are averaged. Pidgeon et al. 2010; American Journal of Science, by permission)

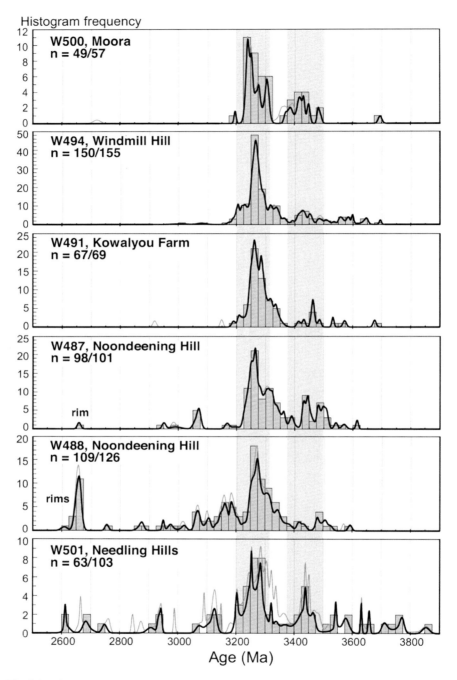

Fig. 9.3 Distribution of detrital zircon ages in the Yilgarn Craton. Probability density diagrams and histograms of ^{207}Pb/^{206}Pb ages obtained for zircons from six quartzite samples in the southwestern Yilgarn Craton. Thick curves and frequency histograms (bin width 25 Ma) include only data <5 % discordant, thin curves include all data. N = ages <5 % discordant/total number of ages (From Pidgeon et al. 2010; American Journal of Science, by permission)

9.1 Evolution of the Yilgarn Craton

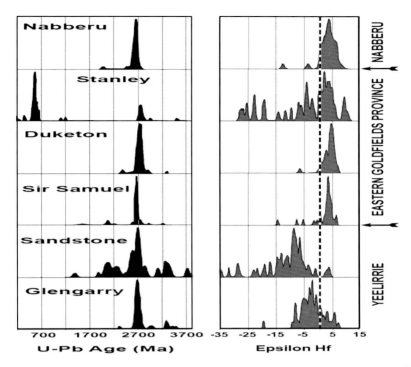

Fig. 9.4 Frequency distribution of isotopic U-Pb zircon ages and of initial εHf indices (^{176}Hf/^{177}Hf)$_{initial}$ in several terrains of the Yilgarn Craton. *Left*: relative probability plots of U–Pb data from each area. *Right*: cumulative probability plots of εHf data from each area, with vertical bar representing CHUR composition (From Griffin et al. 2004; Elsevier by permission)

the Yilgarn Craton including the Murchison and Southern Cross provinces and the high-grade southwest Wheat Belt province contain ~3.0 Ga metavolcanic rocks intruded by granite sheets at 2.9 Ga. By contrast the granite-greenstone terrain of the Eastern Goldfields consists mostly of 2.7–2.6 Ga volcanic sequences and granitic plutons, a period associated with extensive deformation, including deep crustal thrusting and development of imbricated structures (Fig. 9.1).

Geological studies of the Yilgarn Craton are constrained by extensive regolith, weathering and surficial cover (Whittaker 2001). Airborne magnetic data flown at 60 m above ground and 1,500 and 400 m line spacing allow discrimination between gneiss-migmatite-granite, banded gneiss, granite plutons and greenstone domains, defining magnetic/structural zones separated by sharp boundaries, including the Narryer, Murchison, Toodyay-Lake Grace, Southwest, Southern Cross, Yeelirrie, Lake Johnston, Eastern Goldfields and Norseman-Wiluna Belts (Gee et al. 1981; Whittaker 2001) (Fig. 9.1).

Early studies of Yilgarn greenstone belts in the Eastern Goldfields of Western Australia outlined sequences ranging from ultramafic and mafic basalts at low stratigraphic levels to felsic volcanics and turbidites at mid-stratigraphic levels to molasse-like conglomerates at the top, compared at the time to Alpine ophiolite-melange sequences (Glikson 1972a, b). Similar sequences were observed in Canada, South Africa, and India. The low K2O levels of meta-basalts are comparable to oceanic tholeiites (Hallberg 1971). The composition of felsic lavas and hypabyssal dykes and sills is predominantly Na-rich, by analogy to Alpine Na keratophyres. These observations were interpreted in terms of an evolutionary trend from a thin primordial oceanic crust to a geosynclinal pile. However, uniformitarian interpretations have to be modified or abandoned in view of further geochemical and isotopic studies (Chaps. 6 and 12).

Gee et al. (1981) subdivided the Yilgarn Craton into four provinces, the Western Gneiss Terrain, and three granite-greenstone systems to the east, including the Murchison, Southern Cross and the Eastern Goldfields Provinces. U-Pb Zircon studies define ages of 3.7–3.3 Ga for gneiss in the Northwestern Gneiss Terrain, 3.0–2.7 Ga for greenstones in the Murchison and Southern Cross systems, and 2.74–2.63 Ga for greenstones and granite intrusions in the Eastern Goldfields Province. Most terrains contain rare ~3.0 Ga zircon xenocrysts (Swager et al. 1997; Cassidy et al. 2006). Northwestern granites include zircons with Hf-isotope compositions suggesting derivation from source terrains ~3.7 Ga (Wyche 2007). By contrast, no zircons older than ~3.055 Ga have been identified in the Southwestern terrain, the north-central Youanmi terrain and the Eastern Goldfields Super-terrain. In the Youanmi terrain, north-central Yilgarn Craton, eruption of mafic-ultramafic volcanics post-3.1 Ga signifies plume related rift volcanism. The voluminous granites intruded during 2.75–2.62 Ga dismembered and deformed the greenstone depositories, producing the present-day granite-greenstone patterns (Wyche 2007).

Isotopic age frequency distribution plots define distinct peaks representing crustal forming and reworking events, as follows: (1) U-Pb ages of detrital zircons define age peaks at ~3.75–3.6 Ga, 3.4–3.36 Ga, 3.26–3.24 Ga (Fig. 9.2) (Pidgeon and Nemchin 2006; Wyche 2007); (2) zircons derived from quartzites from the southwestern Yilgarn Craton display distinct age peaks between ~3.3 and 3.2 Ga (Fig. 9.3) (Pidgeon et al. 2010); (3) U-Pb data and $\varepsilon Hf_{initial}$ indices indicating concentration of ages about 2.7 Ga display high-$\varepsilon Hf_{initial}$ mantle-type sources in northern (Nabberu) and central (Eastern Goldfields) terrains of the Yilgarn Craton and low- $\varepsilon Hf_{initial}$ crust type sources in the north-central Yeelirrie terrain (Griffin et al. 2004) (Fig. 9.4).

Ion probe (SHRIMP) U–Pb studies of detrital zircons derived from ~3.1 Ga quartz-rich metasediments (Pidgeon et al. 2010) from the Toodyay-Lake Grace Domain, South West Terrain, demonstrate a basic uniformity in the composition of the source rocks derived from a provenance dominated by ca. 3,350–3,200 Ma granitic rocks with an age peak at ~3,265 Ma. Granites of this age have not been identified to date in the southwest Yilgarn Craton but ~3,280 Ma granites occur in the Narryer terrain, northwestern Yilgarn Craton. A second consistent zircon age component suggests an earlier episode of granite emplacement at ~3,500–3,400 Ma and a minor component is identified as 3,850 Ma. The source of zircons in the southwestern Yilgarn Craton is different from that of zircons contained in ~3.1 Ga quartzites and conglomerates from Mt Narryer and the Jack Hills, northwestern Yilgarn Craton, which display a zircon population in the main range of ~3,750–3,550 Ma (Fig. 9.2), and from the Maynard Hills greenstone belt in the Southern Cross Domain which display a zircon population in the range of 3,450–3,250 Ma (Fig. 9.2). The overlap between the peak zircon ages of ~3,220–3,300 Ma (Fig. 9.3) and the large ~3.265–3.225 impact cluster recorded in the Barberton Greenstone Belt and its correlatives in the Pilbara Craton is considered significant (Sects. 7.3 and 8.3).

Van Kranendonk et al. (2013) indicate broad analogies between contemporaneous domains of the Yilgarn Craton, including the Murchison and Eastern Goldfields terrains. The similarities include ~2,810 hypabyssal and volcanic mafic-ultramafic magmatic activity, 2,720 Ma komatiitic-basaltic volcanism followed by widespread 2,690–2,660 Ma felsic magmatism, early deformation at 2,675 Ma, shear-hosted gold mineralization at 2,660–2,630 Ma, and post-tectonic granites at c. 2,630 Ma. These authors suggest that the shared evolutionary sequence, prehnite-pumpellyite to upper greenschist facies, lack of evidence for significant thrusting, and lack of passive margin/foreland basin/accretionary prisms require re-examination of subduction-accretion tectonic models.

Griffin et al. (2004) applied U–Pb age determinations, Hf-isotope and trace element analyses of detrital zircons to assess the geochronology and crustal history of segments of the Yilgarn Craton, including 550 analyses of detrital zircons from the northern Yilgarn Craton (Fig. 9.4). These studies identified ~3.7 Ga zircon ages and similar Hf model ages in younger granites of the Yeelirrie domain. In the Narryer province, northwestern Yilgarn Craton, traces of >3.4 Ga rocks are represented in younger igneous compositions. By contrast Hf-isotope data suggest much of the central and eastern Yilgarn Craton formed later than ~3.2 Ga. The 2.3–1.8 Ga magmatic activities associated with the development of the inter-cratonic Capricorn mobile belt involved recycling of Archaean crust.

Early interpretations of the origin of Archaean cratons included a view of primitive mafic-ultramafic SIMA crust intruded by domal plutons dominated by the tonalite-trondhjemite-granodiorite (TTG) association (Glikson 1972a, b; Glikson and Sheraton 1972; Glikson 1979a, b). Other models invoked development of greenstone belts above SIAL crust, or as intra-cratonic rift zones (Gee et al. 1981; Groves and Batt 1984). Some authors regarded the Norseman-Wiluna greenstone belt as a westward-dipping subduction (Barley et al. 1989; Morris 1993). Campbell and Hill (1988) invoked plume tectonic processes. The stratigraphy of greenstone belts was compared with plate tectonics sequence stratigraphy models, suggesting thin-skinned fold and thrust tectonics and transcurrent movements, consistent with allochtonous sheets truncated at 4–7 km depth by sub-horizontal detachments in contact with underlying SIAL crust. The evidence for large Late Archaean asteroid impact events (~2.63 Ga, ~2.57 Ga, ~2.56 Ga, ~2.48 Ga) observed in the Fortescue Basin and West Transvaal Basin (Sect. 10.1) is relevant to theories of the evolution of greenstone-granite systems (Chapter 12).

9.2 Evolution of South Indian Granite-Greenstone Terrains

The South Indian Shield displays continuous transitions from northern low metamorphic grade greenstone belts to southern gneiss-granulite terrains, allowing reconstruction of the Archaean crust in three dimensions. The shield includes six main cratons (Fig. 9.5; see also Fig. 3.2):

(A) Dharwar craton (also called Karnataka craton);
(B) Bastar craton (also called Bastar-Bhandara craton);
(C) Singhbhum craton (also called Singhbhum-Orissa craton);
(D) Chotanagpur Gneiss Complex;
(E) Rajasthan (Bundelkhand) craton;
(F) Meghalaya craton;

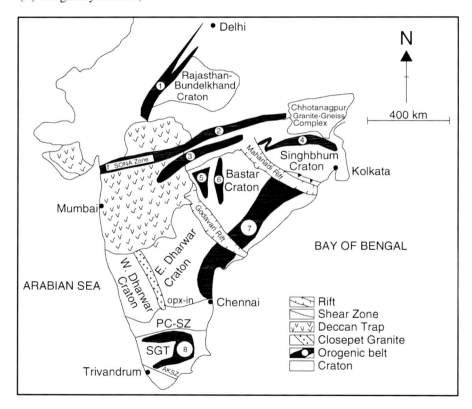

Fig. 9.5 Outline map of the Indian shield showing the distribution of cratons, including Chotanagpur Granite Gneiss Complex, Rifts and Proterozoic fold belts. The fold belts are: *1* = Aravalli Mt. belt, *2* = Mahakoshal fold belt, *3* = Satpura fold belt, *4* = Singhbhum fold belt, *5* = Sakoli fold belt, *6* = Dongargarh fold belt, *7* = Eastern Ghats mobile belt, *8* = Pandyan mobile belt. Abbreviations: *Opx-in* orthopyroxene-in isograd, *AKS* Achankovil shear zone, *PC-SZ* Palghat Cauvery Shear Zone, *SGT* Southern Granulite Terrain, *SONA Zone* Son-Narmada lineament (From Sharma 2009; Springer, by permission)

The Dharwar Craton (Fig. 9.5), extending over almost 500,000 km² in southern India, consists of gneiss and granite-greenstone belts (Ramakrishna and Vaidyanadhan 2008; Sharma et al. 1994), grading into a granulite terrain in the south. The craton is delimited by the upper Cretaceous Deccan plateau basalts in the north, 2.6 Ga-old Karimnagar granulite in the northeast, and the Cuddapah Basin and Proterozoic Eastern Ghats mobile belt in the east (Fig. 9.5). The craton consists of north-northwest trending terrains, separated by the elongated Closepet Granite and by the Chitradurga shear zone (Naqvi and Rogers 1987) (Fig. 9.5). The western belts of the Dharwar Craton (WDC) consist of sediment-dominated supracrustals with lesser volcanic components and were metamorphosed under intermediate-pressure kyanite-sillimanite facies conditions, likely related to associated ~2.6–2.5 Ga granite intrusions. By contrast the eastern supracrustal belts (EDC) are metamorphosed under low-pressure andalusite-sillimanite facies rocks, including fuchsite quartzites, metapelites, andalusite-bearing, corundum-bearing or cordierite-hypersthene-sillimanite-bearing schists and gneisses, calc-silicate rocks and banded manganese-rich BIF. These observations are corroborated by geophysical surveys indicating a thick (40–45 km) nature of the crust underlying the intermediate-pressure WDC and a thinner (35–37 km) crust underlying the EDC (Singh et al. 2003).

The southern part of the WDC includes the 3.3–3.1 Ga greenstone belts of the Sargur Group, as well as younger 2.8–2.6 Ga greenstone belts of the Dharwar Supergroup, separated by angular unconformities. The greenschist to amphibolite facies supracrustal belts include basalts with clastic and chemical sedimentary intercalations. Locally basal conglomerate, cross-rippled quartzite and carbonate sediments occur, consistent with deposition of the Dharwar Supergroup on a gneiss basement (Sharma et al. 1994). The metamorphic grade of the supracrustal belts rises from north to south, reaching orthopyroxene granulite facies of rocks termed *charnockite* associated with orthopyroxene-bearing magmatic rocks termed *enderbite* (Pichamuthu 1960; Ravindra Kumar and Raghavan 1992; Ramiengar et al. 1978), with the metamorphic grade varying locally in relation to associated plutonic bodies. The variations in metamorphic grade along and across the supracrustal belts suggest an overall tilt of the shield eastward as well as northward (Mahadevan 2004). In many areas metamorphic foliation and metamorphic segregation are superposed on and oriented subparallel to original sedimentary layering.

The WDC has been subdivided into Bababudan Group and Chitradurga Group (Maibam et al. 2011). The former consists of a sequence of quartz pebble conglomerate, quartz arenite, basalt, rhyodacite and banded iron formation (BIF). Sm–Nd ages of mafic volcanic rocks indicate volcanic activity at 2.91–2.85 Ga (Anil Kumar et al 1996) whereas SHRIMP U–Pb zircon ages of volcanic tuff interbedded with BIFs yield 2.72 Ga (Trendall et al. 1997). Chitradurga Group assemblages include quartz arenite, carbonate and banded manganese-rich BIF, overlain a basalt-dacite-rhyodacite-greywacke-iron formation sequence. The felsic volcanics were dated by SHRIMP U–Pb zircon as 2.61 Ga (Nutman et al. 1996; Trendall et al. 1997) while mafic volcanics were dated by Sm–Nd isochron age as 2.75 Ga-old (Anil Kumar et al 1996).

Gneisses enveloping the supracrustal belts, broadly denoted as *Peninsular Gneiss*, include a widespread tonalite-trondhjemite-granodiorite (TTG) suite dominated by

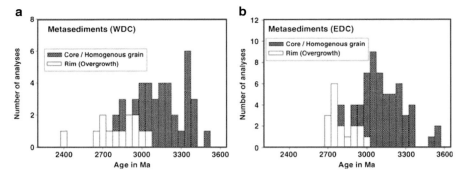

Fig. 9.6 Histograms of age distributions of cores and rims of detrital zircons from the metamorphosed sedimentary rocks of Western Dharwar Craton (*WDC*) and Eastern Dharwar Craton (*EDC*) (Maibam et al. 2011; Springer, by permission): (**a**) Western Dharwar Craton (*WDC*); (**b**) Eastern Dharwar Craton (*EDC*)

3.56–3.4 Ga and ~3.0 Ga tonalitic gneisses which engulf ~3.2–3.0 Ga metamorphosed volcanic and sedimentary rocks denoted as *Sargur Group* (Meen et al. 1992). However, field relations do not always allow distinction between older xenoliths and younger tectonically interleaved supracrustals (Glikson 1979a, b, 1982). A compilation of principal characteristics of the EDC and WDC terrains is presented by Maibam et al. (2011).

Histograms of U-Pb age distribution of detrital zircons in the EDC and WDC supracrustal sediments are shown in Fig. 9.6 (Maibam et al. 2011). Zircons in metasediments, orthogneiss and paragneiss from both the WDC and the EDC yielded ^{207}Pb–^{206}Pb ages of 3.5–3.2 Ga whereas younger age imprints are represented by zircon overgrowth zones, representing evolution of both the EDC and WDC terrains between ~3.5 and 2.5 Ga, consistent with deformation patterns reported by Chadwick et al (2000). These authors drew parallels between the evolution of the Dharwar Craton and modern magmatic arcs, based on the presence of intermediate volcanic assemblages including boninites, andesites, adakites and dacites (Manikyamba et al. 2008). This involved reworking or remobilization of >3 Ga-old crust in the EDC consequent on large-scale arc magmatism. Large-scale K-rich granitic activity in the EDC resulted in higher heat flow in the EDC relative to the WDC. In this model the Chitradurga fault, which separates the EDC and WDC, is regarded as an oblique slip fault between foreland basin region and the magmatic arc region. Evidence for a short-lived nature of the Late Archaean magmatism in the Dharwar Craton is derived from U-Pb ages and Lu-Hf isotope signatures (Mohan et al. 2014). This includes voluminous ~2.56 Ga granitic magmatism in the Western Dharwar Craton as well as previously studied ~2.61 Ga and 2.54–2.52 Ga felsic igneous activity, consistent with several short-lived, episodic crustal growth events over a period of ~100 m.y.

9.3 Accretion of Superior Province Terrains

Goodwin (1968, 1981) proposed an analogy between the progressive accretion of greenstone-granite belts across the Superior Province and circum-Pacific arc-trench accretionary systems, a concept supported by subsequent studies (Langford and Morin 1976; Card 1990; Williams et al. 1992; Stott 1997), leading to models of complex aggregation and accretion of SIAL cratons. Percival et al. (2004, 2006) observe early micro-continent and arc-continent collisions, including the micro-continental terrains of the Northern Superior Province, Hudson Bay and Minnesota River Valley assembled about ~2.72–2.68 Ga. Early ~3.8 Ga rocks occur in the north-western Superior Province (Skulski et al. 2000) and the Hudson Bay terrain in the northeast. By contrast latitudinal-elongated terrains to the south are mostly of mid to late Archaean age, an example being the extensive ~3.0 Ga North Caribou terrain in the south Superior Province (Stott and Corfu 1991; Stott 1997). Early to mid-Archaean ~3.6–3.0 Ga cratonic fragments include the Minnesota River Valley terrain (Goldich and Hedge 1974), Winnipeg River and Marmion terrains. The older cratonic terrains are separated by elongated volcanic belts containing isotopic and trace element evidence for derivation from juvenile mantle sources and a paucity of cratonic clastic sediments, by analogy to Phanerozoic oceanic, arc-trench magmatic subduction domains (Thurston 1994; Kerrich et al. 1999; Percival 2007). Examples include the Oxford–Stull, La Grande, western Wabigoon and Wawa–Abitibi terrains of the southern Superior Province. Extensive latitudinal Late Archaean Large sedimentary belts including greywacke turbidites, migmatite and derived granite, including the English River, Quetico and Pontiac belts, are viewed syn-orogenic flysch to post-tectonic molasses-like wedges representing collisional progenies (Williams et al. 1992; Davies 2002). Seismic studies indicate a northward subduction polarity (White et al. 2003).

The accretion and amalgamation of Archaean micro-continental cratons and juvenile greenstone belts culminated with the locking of the mid to Late Archaean supracrustal and cratonic belts between the ~3.8 and 2.9 Ga Hudson and Northern Superior terrains in the north and the ~3.5–2.8 Ga Minnesota River valley craton in the south. Airborne and magnetic data define a boundary between the ~2.76 and 2.7 Ga Wawa granite-greenstone terrain and the ~3.5–2.8 Ga Minnesota River gneisses, referred to as the Great Lakes Tectonic Zone (Morey and Sims 1976). The mid-Archaean gneisses represent tectonically interleaved metamorphosed plutonic and supracrustal rocks of a range of ages (Bickford et al. 2006) including significant tectono-thermal events at ~3.5 Ga and 3.38 Ga and synkinematic and post-kinematic thermal events at ~2.6 Ga. The Superior Province greenstone-granite terrain thus likely represents an Archaean analogue of Phanerozoic plate tectonic-related transformation and accretion of SIMA to SIAL crust.

9.4 Archaean Fennoscandian/Baltic Terrains

Supracrustal suites of the Archaean of the Fennoscandian (Baltic) Shield are comprised of a number of age components, including <3.2 Ga, 3.1–2.9 Ga, 2.90–2.82 Ga, 2.82–2.75 Ga and 2.75–2.65 Ga, including granitoid complexes, ophiolites and eclogite-bearing associations (Slabunov et al. 2006). The main part of the shield with ages in excess of 3.0 Ga consist of the Karelia granite-greenstone province. Whereas detrital zircon grains of early Archaean age are present, the bulk of the Fennoscandian Shield consist of 3.2–3.1 Ga rocks, signifying the emergence of the first micro-continents such as Vodlozero and Iisalmi, which reached a peak about 2.90–2.65 Ga ago. Most authors advocate analogies with Phanerozoic ensialic and ensimatic subduction-related systems, including collisional, spreading-related, continental rifting, and the setting related to mantle plumes (Slabunov et al. 2006). East of the Fennoscandian Shield a continuous transition occurs to Proterozoic system of paleo-rift structures of the East European platform. Rifting and break-up of the shield during 2.5–2.0 Ga dispersed the Archaean domain (Gorbatschev and Bogdanova 1993). In the northern part of the shield basins underwent collisional orogeny during the Svecofennian orogeny 1.95–1.82 Ga-ago, expanding the continental crust. In the west the Svecofennian Orogen is truncated by the ~1.8–1.65 Ga ensialic Trans-Scandinavian igneous belt. Final stages of SIAL formation occurred in western Scandinavia between ~1.75 and 1.55 Ga, including formation of anatectic en-sialic granitoids. This was followed by major en-sialic reworking during the Sveco-Norwegian-Grenvillian and Caledonian orogenies ~1.2–0.9 and 0.5–0.4 Ga ago, respectively.

Chapter 10
Post-3.2 Ga Basins and Asteroid Impact Units

Abstract The sedimentary records of pristine little-deformed upper to late Archaean basins, including the Fortescue Basin, Hamersley Basin and West Transvaal basin, contain a number of major asteroid impact ejecta/fallout units spanning ~2.63–2.48 Ga. Detailed field and laboratory studies by Bruce Simonson, Scott Hassler and others uncovered a wealth of detailed evidence for the nature of the impacts and related mega-tsunamis. Follow-up platinum group element (PGE) analyses help to resolve the composition and size of the projectiles. The location of the original impact sites from which the ejecta was derived remains unknown. Stratigraphic and isotopic age studies define a number of inter-continental correlations between impact ejecta horizons in the Hamersley and Transvaal basins. These impact events bear implications for the origin of contemporaneous late Archaean greenstone-granite systems. Possible relations between large impacts and enrichment of the sea water in iron are subject to current studies, with potential implications for the origin of banded iron formations.

Keywords Upper Archaean • Fortescue Group • Hamersley Group • Transvaal Basin • Asteroid impacts • Microkrystite spherules • Tsunami breccia

10.1 Late Archaean Griqualand West Basin

The Griqualand West Basin of South Africa (Figs. 10.1 and 10.2) displays a wealth of primary sedimentological, palaeo-biological and asteroid impact fallout features (Button 1976; Cheney 1996; Beukes and Gutzmer 2008; Simonson et al. 2009a, b). The Monteville Formation (2,650 ± 8 Ma), consisting largely of carbonates and shale, includes an impact spherule layer (Simonson et al. 1999). Overlying carbonates of the Reivilo Formation host a younger spherule layer and a third spherule layer occurs in a core of the Kuruman Banded Iron Formation (Simonson et al. 2009a, b).

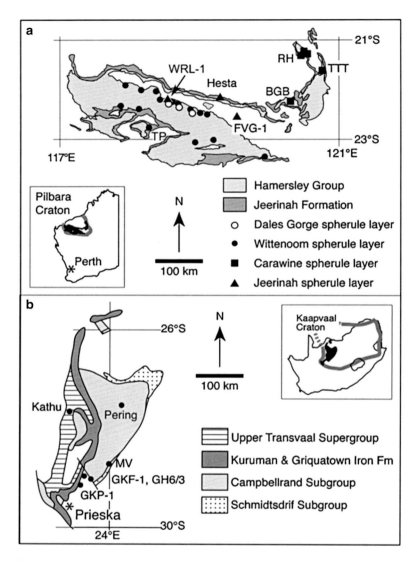

Fig. 10.1 (a) Hamersley Basin map showing areas of occurrence of Hamersley Group and Jeerinah Formation (*uppermost Fortescue Group*) and most locations where spherule layers have been studied; labeled locations are Billy Goat Bore core drill site (*BGB*), two CRA core drill sites (FVG-1 and WRL-1), Hesta exposure, Ripon Hills exposures (*RH*), Tarra Tarra turnoff exposure (*TTT*), and Tom Price (*TP*). See Hassler et al. (2005) and references therein for more details; (b) Griqualand West Basin map showing areas of occurrence of units in the Transvaal succession labeled are Agouron core drill sites (GKF-1 and GKP-1), an exploration core drill site nearby (GH6/3), Kathu core drill site, Monteville exposures (*MV*), and Pering mine (From Simonson et al. 2009a, b; Elsevier, by permission)

10.1 Late Archaean Griqualand West Basin

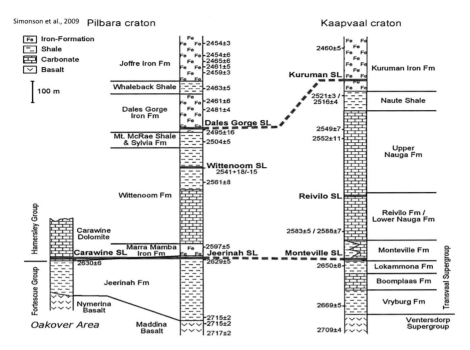

Fig. 10.2 Schematic columns of the parts of the Hamersley and Griqualand West successions on the Pilbara and Kaapvaal cratons, respectively, showing known impact spherule layers and proposed correlations. The middle column represents the stratigraphy throughout the main Hamersley Basin, whereas the short column on the left represents the erosional truncated succession typical of the Oakover River area in the northeast, e.g., the Ripon Hills area. Numbers on the side of each column list age constraints in millions of years. Dashed lines indicate layers thought to have been produced by the same impact (Simonson et al. 2009; Elsevier, by permission)

2,650 ± 8 Ma. Monteville Spherule Layer (MSL). Simonson et al. (2000a, b) reported a later Archaean microkrystite layer approximately ~8 cm-thick from the sedimentary tuff, carbonate and shale-dominated Monteville Formation, present in ten cores distributed throughout an area of ~17,000 km². The MSL spherules are dominated by partly carbonate-altered K-feldspar typical of microkrystite spherules (Smit and Klaver 1982) and compositions, textures, and sedimentary structures typical of impact fallout units, including high Ni, Co and PGE, chondrite-like distribution patterns of the PGE and Ni/Ir, with Ir levels up to 6.4 ppb, as compared to mean crustal levels of ~0.02 ppb and chondritic levels of ~500 ppb. The Monteville spherule layer is likely correlated with the ~2.63 Ga Jeerinah Impact Layer of the Fortescue Basin, Pilbara Craton, Western Australia (Figs. 10.1 and 10.2).

2,581.9 Ma. The Reivilo spherule layer (RSL), first identified in the Kathu core and three other cores of the Griqualand West Basin (Simonson et al. 1999), consists of shallow sub-tidal carbonate (Sumner and Beukes 2006; Simonson et al. 2009a, b) containing spherule units up to about 2–9 cm-thick and lenses hosted by deep water sediments. Spherules consist of mm-scale microkrytite spheres composed largely of K-feldspar, sericite, pyrite and carbonate (Simonson and Carney 1999). Spherule textures are in the main comparable to those of the K–T boundary layer (Izett 1990; Bohor and Glass 1995). Some spherules are tear-drop shaped suggestive of aerodynamic effects. Internal textural features of the spherules include pseudomorphs of 5–50 µm laths, K-feldspar crystallites ~5–700 µm-long, deformed crystallites, radiating fans, inward radiating fans branching from outer spherule rims, and 0.06–1 mm large central spots filled with carbonate, K-feldspar or sericite and representing original melt cavities. Simonson et al. (1999) regard the K-feldspar as authigenic. Spherules are commonly compacted although internal textures are mostly preserved intact, suggesting hard-shell protection upon deposition, perhaps somewhat analogous to the much larger-scale pillow lavas. Broken spherules are present. Spherule layers may be overlain by rip-up clasts zones up to 20 cm thick, suggestive of tsunami effects. Host carbonate sediments of the RSL may include small stromatolites, indicating shallow water environment. The Reivilo spherule layer may correlate with the ~2.56 Ga Spherule Marker Bed in the Bee Gorge member of the Wittenoom Formation (Simonson 1992; Glikson 2004a, b; Glikson and Vickers 2007) or with the ~2.57 Ga Paraburdoo spherule Layer, Hamersley Basin, Pilbara Craton, Western Australia (Hassler et al. 2011). Laser-ICPMS study of Reivilo spherules has detected Ni-Fe particles and Ni values of 3,931 ppm (Goderis et al., 2013).

<2,521 ± 3 Ma. The Kuruman spherule layer (KSL), identified in the Agouron drill core (Simonson et al. 2009a, b) where it is located approximately ~37 m above the base of the Kuruman Iron Formation, constitutes a <98 cm-thick layer of graded coarse-grained spherules consisting mainly of stilpnomelane and carbonate. The spherules occur mostly at the base of the layer and are overlain by compacted rip-up clasts of carbonate, chert, and/or stilpnomelane. Spherules are commonly rimmed by K-feldspar, which also occur as inward radiating fans and internal crystallites. Internal zones of stilpnomelane contain skeletal pseudomorphs after unidentified original phases. Rip-up clasts are about 3–5 mm long and 1 mm thick and some significantly larger. A possible upper 6 mm-thick layer of spherules dominated by stilpnomelane but devoid of K-feldspar, and thus compositionally distinct from the RSL, occurs at the top of the Kuruman Iron Formation in the GKF-1 drill core (Simonson et al. 2009a, b). The layer, containing rip-up clasts of stilpnomelane and carbonate up to 7 mm or more large, is overlain by an anomalous 17 mm layer of carbonate and sericite.

Siderophile trace element patterns of spherules from the Griqualand West Basin are consistent with the presence of a meteoritic component, including high total PGE levels and low Pd/Ir ratios (Table 10.1). The low Pd/Ir levels relative to terrestrial basalts can only be produced by primary chondritic Pd/Ir ratios of ~1.2 to which Pd – a mobile element in the terrestrial environment has been added during diagenetic processes. The only alternative, strong fractionation of the PGE represented by low Pd/Ir of harzburgites residues (Chou 1978), is unlikely to apply to the spherule beds.

Table 10.1 Siderophile and PGE trace element comparisons between the mean composition of spherule units from the Griqualand West Basin

Monteville	Ni ppm	Cr ppm	Ni/Cr	Ir ppb	Pd ppb	Pd/Ir	Ref
Spherule bed							
P9-1A	65	418	0.15	4.8			1
P11-1A	15	385	0.038	2.5			1
P11-1B	68	264	0.25	6.4			1
P11-1B1	167	180	0.93	5.6			1
P11-2B	16	11.9	1.3	0.15			1
TK-1A	39.8	340	0.12	2.57	7.46	2.9	1
U98-1B	54.3	67	0.81	0.62	1.38	2.2	1
U126-1	27.2	39	0.69	0.28	0.67	2.4	1
V-1	35.7	293	0.12	4.92	6.85	1.4	1
Reivilo spherule bed							
Ku32sl	3,931	830	4.73	176	59.4	0.34	
A – C1	10,500	2650	4	455	550	1.2	
B – pyrolite	1,960	2625	0.75	3.2	3.9	1.2	
C – mean ultramafic	2,000			2.4	8.2	3.4	
D – komatiite	1,530			1.0	8.4	8.4	
E – MORB	200			0.05	0.6	12	
F – Archaean tholeiite	<150	<100	~1.5				
G – Archaean -high-Mg	198	749	0.26	0.49	18.2	37	

Simonson et al. (2009a, b) [Ref 1], C1-chondrites (A – Chou 1978); mantle pyrolite (B – McDonough and Sun 1995); mean ultramafic (C – Chou 1978); komatiite (D – Chou 1978); mid-ocean ridge basalt (E – Chou 1978); average Archaean tholeiite (F – BVSP 1981); high-Mg basalt (G – Glikson and Hickman 1981 – mean of 20 high-Mg basalts of the Apex Basalt)

10.2 Late Archaean Fortescue and Hamersley Basins Impact Units

10.2.1 ~2.63 Ga Jeerinah Impact Layer and the Roy Hill and Carawine Dolomite Impact/Tsunami Megabreccia

Simonson et al. (2000a, b) reported a thin ~6 mm layer of microkrystite spherules from the top of a black shale sequence of the Jeerinah Formation in the core of drill hole FVG-1 at Ilbiana Well, Fortescue River valley. The layer, denoted as Jeerinah Impact Layer (JIL), occurs 2.7 m below the top of the Jeerinah Formation (2,684 ± 6 Ma–2,629 ± 5 Ma) (Arndt et al. 1991; Nelson 1999) underlying the Marra Mamba Iron Formation (2,597 ± 5 Ma). An exposure of the top-Jeerinah Formation impact spherule unit was located by B.M. Simonson at Hesta siding about 50 km northwest of Ilbiana Well (Simonson et al. 2001) (Figs. 10.1 to 10.8). This locality exposes the full transition from laminated argillite to argillite-chert sequence (~5 m) to a ~1 m thick unit consisting of (1) a lower zone of microkrystite spherules-bearing breccia with siltstone and chert intraclasts (~40 cm); (2) a massive zone of spherules with fewer argillite and chert intraclasts (~60 cm), and (3) overlying

Fig. 10.3 ~2.63 Ga Jeerinah Impact Layer (*JIL*) outcrop at Hesta: (**a**) Exposure of section from the Jeerinah Formation siltstones (*JS*) to siltstone-chert (*SC*) to laterite top (*L*). The JIL is located below the laterite cap; (**b**) Boulder bed representing a tsunami overlying the JIL; (**c**) Basal spherule-bearing siltstone breccia of the JIL; (**d**) Spherule-bearing siltlstone breccia containing tektites (*T*); (**e**) Microkrystite spherule consisting of inward-radiating K-feldspar microlites and an offset vesicle; (**f**) Vesicle-bearing microtektite fragment

microkrystite spherules-poor breccia dominated by angular to rounded chert cobbles and boulders reaching 50 cm in size. The impact fallout unit is capped by a ~60 cm-thick argillite, in turn overlain by a boulder deposit likely representing tsunami effects (Figs. 10.5 and 10.6).

The basal conglomerate of the JIL unit consists of a poorly sorted assemblage of rip-up fragments of chert and siltstone set in a matrix of siltstone rich in microkrystite spherules and microtektites. The fragments are of similar composition to the underlying chert-argillite sequence of the Jeerinah Formation. The overlying spherule beds display high concentration of microkrystite spherules and lesser proportion of microtektites set in an argillite matrix. Microkrystite spherules display inward radiating fibrous K-feldspars and centrally offset vesicles and have a size range

10.2 Late Archaean Fortescue and Hamersley Basins Impact Units

Fig. 10.4 ~2.63 Ga Jeerinah Impact Layer (*JIL*). A specimen of microkrystite spherules containing fragments of ferruginous siltstone. Hesta locality, central Pilbara Craton (Courtesy Bruce Simonson)

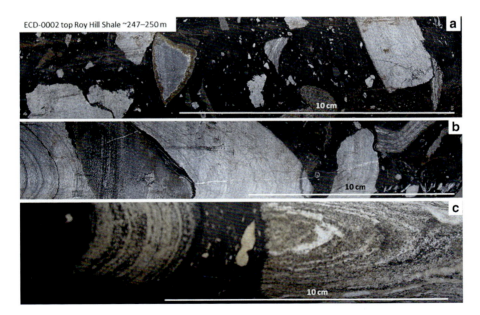

Fig. 10.5 ~2.63 Ga Roy Hill breccia: Chert, carbonate and sulphide fragments and boulders within carbonate-spotted black carbonaceous shale within the tsunami deposit incorporating the Jeerinah Impact layer (*JIL*) in the ECD-0002 drill core, underlying the Roy Hill Member of the Marra Mamba Iron Formation. Note the sulphide rims and blebs in frames A and C

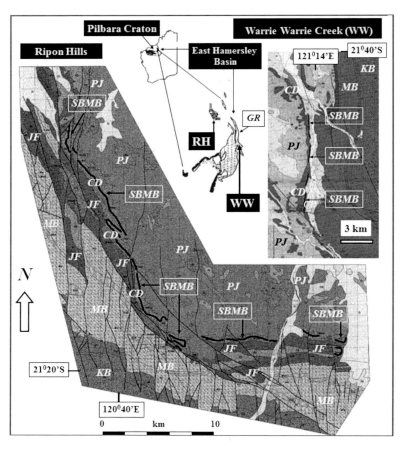

Fig. 10.6 A geological sketch map of the Eastern Hamersley Basin, indicating the distribution of the Carawine Dolomite Megabreccia (CDMB). *KB* Kylena Basalt, *MB* Maddina Basalt, *JF* Jeerinah Formation 2.629 Ga, *PJ* Pinjian Chert (silicified breccia), *GR* Gregory Range, *WW* Warrie Warrie Creek belt, *RH* Ripon Hills. The CDMB unit is shown as a thick black line (Modified after Williams (2003). From Glikson 2004a, b; Astrobiology, by permission)

of 0.3–1.2 mm (Figs. 10.3 and 10.4). Composite grains with small microkrystite spherules enveloped by larger spherules and agglutinated spherules are present. Irregular-shaped fragments consist of feldspar microlites in microcrystalline matrix, showing some flow banding and quartz-filled vesicles, but little or no radial crystal textures. Both types of particles consist mainly of K-feldspar, hydrated iron oxide, sericite, quartz and carbonate. The irregular particles can be compared with Muong Nong type microtektites such as described from tsunami-disrupted carbonate-chert unit located at the contemporaneous Carawine Dolomite (Simonson et al. 2000a, b, 2001; Glikson 2004a, b). Most microtektites, which can reach 2–4 mm in size, are of irregular shape and contain abundant quartz-filled micro-vesicles/bubbles on a scale of <100 μm set in clouded microcrystalline matrix interpreted as meta-glass from its palimpsest flow banding.

10.2 Late Archaean Fortescue and Hamersley Basins Impact Units

Fig. 10.7 Carawine megabreccia (*CDMB*) (**a**) outcrops of CDMB at Rippon Hills, Oakover Valley, eastern Pilbara; (**b**) Tsunami breccia – fragments of chert in carbonate matrix; (**c**) fragment of black chert in carbonate matrix; (**d**) Microtektite (*T*) fragment and composite microkrystite spherule (*MK*) within matrix of breccia (**d** – Courtesy Bruce Simonson)

Laser ICPMS analyses of JIL microkrystites indicate Ni levels are similar to those of basaltic compositions, Ni/Co ratios are somewhat higher than those of basaltic compositions. Pd/Ir ratios in JIL and CDMB spherules are distinctly low (~0.8–2.1) as compared with those of basalts and komatiites (~3.4–12) and similar relations hold for Pt/Ir (Table 10.2), consistent with earlier observations regarding depletion of volatile PGE relative to refractory PGE in microkrystite spherules during condensation (Glikson and Allen 2004; Glikson 2007).

The occurrence of <5.0 mm-large microtektites and fragments of meta-glass in the JIL may militate for a relative proximity of the originating impact crater, namely in the order of no more than a couple of thousand kilometer from the impact fallout location. Alternatively, the incorporation of large melt fragments suggests the impact was of a magnitude resulting in global fallout of ejecta. The large sizes of some of the spherules, reaching above 1 mm at the basal spherule-intraclast unit, may correlate with impacts by asteroids on the scale of 10–20 km following the criteria of O'Keefe and Ahrens (1982) and Melosh and Vickery (1991). Impacts by asteroids on this scale would result in impact craters with diameters in the order of 150–300 km. The prevalence of chlorite in spherule interiors and the lack of shocked

Fig. 10.8 Carawine Megabreccia (CDMB): (**a**) conformable breccia (labelled B) intercalated between horizontal slabs of Carawine Dolomite (*CD*) and vein breccia (*arrows*) injected across layering; (**b**) fragments of chert within dolomite; (**c**) microkrystite spherule overlying the CDMB, displaying inward-radiating K-feldspar surrounding a core of quartz and Fe-oxide; (**d**) spherule consisting of chlorite shell enveloping carbonate-chlorite core

quartz grains with planar deformation features are suggestive of a basaltic/oceanic composition of the target crust, by analogy to earlier observations pertaining to microkrystites of the Wittenoom Formation and the Carawine Dolomite.

The development of JIL at Hesta is interpreted in terms of the following stages:

(A) Impact-triggered seismic perturbations represented by rip-up clasts forming intraclast breccia concomitant with settling of microkrystite spherules, microtektites and microtektite fragments.
(B) Continuing deposition of microkrystite spherules, microtektites and meta-glass fragments associated with minor rip-up clasts.
(C) Deposition of tsunami wave-transported cobbles and boulders of chert.
(D) Return to below-wave base deposition of argillite.

Rasmussen and Koeberl (2004) report an Ir concentration of 15.5 ppb in JIL spherules and an angular quartz grain containing planar deformation features (PDF),

10.2 Late Archaean Fortescue and Hamersley Basins Impact Units

Table 10.2 Siderophile element abundances of spherule-bearing sediments from the JIL and CDMB units

	Ni ppm	Co ppm	Cr ppm	Ni/Co	Ni/Cr	Ir ppb	Pt ppb	Pd ppb	Pd/Ir	Pt/Ir	Pd/Pt	Ref
CDMB Av of 5 samples	62	43	22	1.44	2.82	0.596	5.82	1.24	2.1	9.8	0.21	1
W94-1A	176	19.2	251	9.2	0.7	5.18	10.1	7.94	1.5	1.9	0.78	2
W94-1 N	194	17.7	327	11	0.59	11.8	22.1	9.12	0.8	1.9	0.41	2
W94-1O	139	16.7	318	8.3	0.44	8.77	18.8	9.98	1.1	2.1	0.55	2
W102-1A	203	34.1	293	6	0.69	9.11	18.4	11.6	1.3	2.0	0.63	2
A – C1	10,500	500	2650	21	4	455	1010	550	1.2	2.2	0.54	
B – pyrolite	1,960	105	2625	18.5	0.75	3.2	7.1	3.9	1.2	2.2	0.55	
C – mean ultramafic	2,000	110		18.2		2.4	10	8.2	3.4	4.2	0.82	
D – komatiite	1,530	104		14.7		1.0	9.3	8.4	8.4	9.3	0.9	
E – MORB	200	41		4.9		0.05	2.3	0.6	12	46	0.26	
F – Archaean tholeiite	<150	60-80	<100	1.8-2.5	~1.5							
G – Archaean -high-Mg	198	49	749	4	0.26	0.49		18.2	37			

Data from Simonson et al. (1998) [Ref 1] and Simonson et al. (2009a, b) [Ref 2]) compared to C1-chondrites (A – Chou 1978); mantle pyrolite (B – McDonough and Sun 1995); mean ultramafic (C – Chou 1978); komatiite (D – Chou 1978); mid-ocean ridge basalt (E – Chou 1978); average Archaean tholeiite (F – BVSP, 1981); high-Mg basalt (G – Glikson and Hickman 1981 – mean of 20 high-Mg basalts of the Apex Basalt)

indicating the impact affected quartz-bearing rocks. Geochemical estimates by these authors suggest the spherule layer comprises about 2–3 wt % chondritic meteorite component.

A unit of impact spherules correlated with the ~2.63 Ga Jeerinah Impact layer (JIL) is observed within a more than 100 m-thick fragmental-intraclast breccia pile in drill cores near Roy Hill, Eastern Hamersley Basin (Hurst et al. 2013) (Fig. 10.5). The sequence represents significant thickening of the impact/tsunami unit relative to the JIL type section at Hesta, central Pilbara, as well as relative to the 20–30 m-thick ~2.63 Ga Carawine Dolomite spherule bearing mega-breccia (CDMB), Oakover River to the East. The location of the JIL and associated breccia under supergene iron ore within the Marra mamba Iron Formation at Roy Hill renders the impact products of economic interest.

A stratigraphically consistent microtektite and microkrystite spherule-rich chert-carbonate megabreccia unit (CDMB), identified in the Oakover Valley, Eastern Pilbara, at Ripon Hills (Simonson 1992) and Woodie Woodie (Hassler et al. 2000), extends over a strike distance of 98 km (Figs. 10.6 to 10.8). The megabreccia unit is located within the lower basinal facies of the Carawine Dolomite, about 30–100 m above the basal contact with the underlying siltstones of the Jeerinah Formation. The megabreccia consists of chaotic unsorted and randomly oriented chert and dolomite clasts and mega-clasts on scales ranging up to several meters and derived from an autochtonous to sub-autochtonous lithologically distinct chert-dolomite unit. Chert bedding segments may occur at high angles to the overall boundary of the CDMB or may show a crude subparallel imbrication (Fig. 10.8). Ductile deformation of bed segments is observed on a range of scales, for example in layered chert-carbonate blocks, within individual boulders and as small folded concretions.

Type outcrops of the CDMB display multiple veins of breccia and microbreccia injected across and along bedding planes of carbonates underlying the megabreccia (Figs. 10.7 and 10.8). Significantly the veins contain microkrystites and microtektites whose preservation despite expected corrosion through friction is interpreted in terms of tsunami-induced hydraulic pressures, minimizing mechanical grinding effects. Microtektite and microkrystite concentrations vary across the breccia and are generally concentrated toward the top, where the breccia is conformably overlain in sharp contact by well layered carbonate.

Salient features relevant to the origin of the CDMB include the following:

1. The CDMB megabreccia represents mega-tsunami effects associated with an extraterrestrial impact event, evidenced by the occurrence of microkrystite spherules and microtektites.
2. The CDMB megabreccia forms a stratigraphically unique time/event marker horizon, allowing the identification of lateral and vertical facies and thickness controls in the Carawine Dolomite.
3. The dominantly irregular structure, unsorted fragments and near-random orientation of blocks and fragments within the CDMB militates for autochtonous to near-autochtonous derivation of the fragments.
4. The absence of stromatolite-facies fragments within the megabreccia is consistent with an autochtonous nature of the brecciation.
5. The size of blocks, observed up to <7 m, and the common excavation of the base of the CDMB provide evidence for high-energy disruption of the sea bed.
6. The location of the CDMB in below-wave-base carbonate-siltstone units suggests the disruption originated by waves of larger amplitude than those resulting from wind-driven or gravity driven currents.
7. Any model regarding the origin of the CDMB must take into account the constraints imposed by the near-perfect preservation of microkrystite spherules within the megabreccia, including injected microbreccia veins and tongues.

Two models for the origin of the CDMB are discussed, including (a) impact-triggered seismic faulting and (b) mega-tsunami effects:

Model A – Impact-triggered seismic faulting

Submarine faulting associated with major seismic disturbances is, in principle, capable of producing block structures and escarpments shedding large blocks and fragments, which can then be transported by gravity currents and/or tsunami downslope. Such features have been described in the vicinity of the Chicxulub impact structure at Belize, Mexico (Pope et al. 1997). Sharp variations in the thickness of the CDMB where megabreccia sections about 12 m thick are juxtaposed with megabreccia sections about 4 m-thick, support a faulting model. However, the results of such faulting can be expected to be heterogeneous, as contrasted with the extension of the CDMB with broadly similar thickness (4–25 m) over large areas (Fig. 10.6). Had faulting been a significant factor, boulders and fragments derived from fault scarps and transported downslope as debris flow would be partly rounded, which is not the case. In most observed outcrops angular blocks and large detached bedding segments of the megabreccia rest either conformably on, or are torn from, the underlying excavated carbonates, with little or no extension of faults into the substratum. Mechanical grinding of breccia fragment can be expected to have resulted in partial to total destruction of the microtektite spherules. Thus whereas locally faulting constituted a factor, on the whole the main features of the CDMB are not consistent with those of a fault-derived breccia.

Model B – Impact-triggered mega-tsunami autochtonous to sub-autochtonous breccia

Simonson (1992), Simonson and Hassler (1997), Hassler et al. (2000) and Hassler and Simonson (2001) proposed a tsunami origin for the Carawine Dolomite mega-breccia and for the Spherule Marker Bed (SMB). In terms of this model, propagation of a major tsunami wave/s occurred from NE to SW, consequent on an impact by a ~5 km-diameter projectile producing a ~60 km-diameter crater located east and north of area occupied at present by the Pilbara Block. The chaotic sub-autochtonous to autochtonous nature of the CDMB is amenable for interpretation in terms of disruption of the sea bed by a high-amplitude tsunami wave. An autochtonous origin, which accounts for the scarcity of round boulders, cobbles or pebbles, is consistent with the observed excavation of the sea floor, and accounts for the preservation of microkrystite spherules due to extreme hydraulic pressure imparted by the tsunami wave, exceeding lithostatic pressure and keeping breccia fragments and spherules separate within a fluid medium, thus preventing corrosion of spherules. The following sequence of events is envisaged in terms of a tsunami model:

1. Arrival of the tsunami with consequent dispersal of the soft sediment column as a mud cloud in the submarine environment. Deformation of ductile sediments is evidenced by occurrence of ductile deformed/folded layers and fragments.
2. Excavation, fracturing and chaotic disruption of the solid substratum below the soft sedimentary column;
3. Settling of condensed microkrystite spherules and microtektite glass fragments from the impact-released cloud, forming a spherule layer
4. Continuing seismic-triggered tsunami waves and injection of parts of the subaqueous mud cloud as viscous fluid under the pressure of the tsunami, resulting in preservation of the microkrystite spherules in veins
5. Settling of the bulk of the mud cloud above the megabreccia.

10.2.2 ~2.57 Ga Paraburdoo Impact Spherule Unit (PSL)

The Paraburdoo Member of the Wittenoom Formation, which consists of thin-bedded dolomite deposited in a deep shelf off-platform environment (Simonson et al. 1993), contains a <2 cm-thick spherule layer (Paraburdoo Spherule Layer – PSL) located ~55 m below the top of the member and ~86 m below a marker tuff horizon dated as 2,561 + 8 Ma (Hassler et al. 2011). A second tuff horizon located hundreds of meters below the PSL has been dated at 2,597 + 5 Ma (Trendall et al. 1998). Based on an estimate of sedimentation rates an approximate age of 2.57 Ga is suggested for the PSL (Hassler et al. 2011). The contacts with underlying and overlying dolomite are sharp and carbonates above and below the PSL are of similar composition and layering styles. The PSL consist entirely of tight packed spherules accompanied with ~4 % intergranular carbonate cement and no extra-basinal detritus was observed. The spherules are internally recrystallized. K-feldspar crystallites form ~60–90 % and are either randomly oriented or form branching clusters and skeletal textures (Fig. 10.11). About two thirds of the spherules contain phlogopite-type skeletal crystals and pseudomorphs (Hassler et al. 2011). These authors interpret the K-feldspar crystallites as alteration products of quench crystallized plagioclase crystallites, by comparison to K-feldspar replacement of plagioclase in mafic Hamersley tuffs (Hassler 1993). Phlogopite pseudomorphs are possibly derived by alteration of quench crystallized olivine and pyroxene crystallites. Absence of current reworking in the PSL suggests below-wave deposition. Hassler et al. (2011) correlate the PSL with the Reivilo Spherule Layer (RSL), Transvaal Basin (Figs. 10.1 and 10.2), with which it shares textural and mineralogical characteristics, including phlogopite crystallites, not seen in other spherule layers. However, the RSL includes rip-up clasts and has likely formed in shallower water environment. Laser-ICPMS analyses of Paraburdoo spherules determined Ni levels of up to 404 ppm, Ir levels of up to 357 ppb and low Pd/Ir ratios in the range of 0.1–0.67 (Goderis et al. 2013).

10.2.3 ~2.56 Ga Spherule Marker Bed (SMB)

The Spherule Marker Bed (SMB), discovered by Simonson (1992) at Wittenoom Gorge, constitutes a unique microkrystite and microtektite-bearing turbidite unit about 0.5–1.0 m-thick within the ~230 m-thick siltstone-carbonate Bee Gorge Member, upper Wittenoom Formation (Figs. 10.9 and 10.10). The SMB is found to extend for at least 324 km E-W throughout the Hamersley Basin, over an area ~16,000 km^2, displaying thickness and facies variations from centimeter-scale layers and discontinuous lenses of densely packed spherules in carbonate-argillite matrix to several decimeter-thick turbidite units with or without dispersed spherules. Observed thicknesses of the SMB turbidite unit are in the 12–130 cm range, and spherule-rich bands and lenses are in the 0.3–6.5 cm range (Simonson 1992).

10.2 Late Archaean Fortescue and Hamersley Basins Impact Units

Fig. 10.9 ~2.56 Ga microkrystite spherule marker bed (*SMB*), showing two impact cycles, SMB-1 and SMB-2, overlain and underlain by carbonate, siltstone, and chert. Each cycle includes a basal layer or series of lenses of microkrystite spherules (*MKZ*) overlain by rhythmic turbidites (seismic zone, SZ), overlain by a cross rippled tsunami zone (*TZ*). The two cycles are separated by a stratigraphically consistent layer of silicified black siltstone denoted as a 'Quiet Zone' (**a** – Munjina Gorge; **b** – Wittenoom Gorge)

The thickness of spherules generally decreases from north to south, but no geographic trend is observed in the thickness of the turbidite (Simonson 1992). Sedimentological studies of 15 sections of the SMB (Simonson 1992; Hassler et al. 2000; Hassler and Simonson 2001) suggest a slope or deep shelf environment, persisting below and above the impact fallout layer. Cross-ripples form consecutive trains of connected bedforms up to 150 m-long, or discontinuous cross-layered lenses. Measurements of 26 sets of symmetrical cross-ripple and climbing-ripples (wavelength 23–40 cm; amplitude 1.0–3.5 cm; average dimension – 32 × 1.5 cm) indicate prominence of southward-directed palaeocurrents (Hassler et al. 2000).

A tsunami origin of the SMB is defined by the close association of these bedforms with microkrystite and microtektite-bearing units and by the below-wave-base position of the SMB, indicating a wave system of amplitude/energy in excess of ordinary atmospheric (wind)-driven wave systems (Hassler et al. 2000). U-Pb zircon ages for a crystal tuff unit located about 75 m below the SMB were reported by Hassler (1991) as 2,603 ± 7 Ma and by Trendall et al. (1998) as 2,561 ± 8 Ma or 2,565 ± 9 Ma (Trendall et al. 2004), the ~2.6 Ga age probably representing older xenocrysts. Woodhead et al. (1998) measured a Pb-Pb carbonate whole rock age of 2,541 + 18/−15 Ma on the

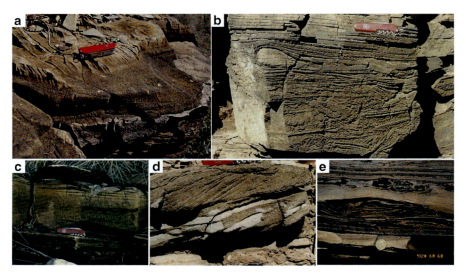

Fig. 10.10 ~2.56 Spherule Marker Bed, Wittenoom Gorge, tsunami-generated structures overlying microkrystite spherule beds: (**a**) <15 cm-thick spherule bed (deep brown dotted bed below swiss knife); (**b**) Development of diamictite structures in differentially siliceous carbonate overlying microkrystite spherules; (**c**) calcarenite-black siltstone intercalations overlying microkrystite spherules and capped by cross-layered tsunami unit' (**d**) cross-layered spherule-rich (more than 80 % spherules) unit; (**e**) cross-cutting sedimentary flame structures overlying spherule unit

SMB, nearly consistent with the U-Pb zircon age. Pb-Pb isotopic analyses of carbonate from the Paraburdoo Member and the Bee Gorge Member yielded ages of 2,505 ± 37 Ma and 2,346 ± 38 Ma, respectively, representing diagenetic overprinting.

Two distinct spherule units are documented within the SMB, each overlain by turbidite and current-disrupted arenite (Glikson 2004a, b) (Fig. 10.9):

SMB-1. The stratigraphically lower cycle contains at its base a cm-scale spherule-rich horizon or discontinuous <5 cm-thick spherule-rich lenses. In many areas spherules are missing. The spherule unit is overlain by graded bedded Bouma-type cycle arenite-turbidites in turn overlain by cross layered arenite, or by siltstone followed by convolute current turbulence climbing cross-bedding and "ball-and-pillow" structures. Major recumbent over-folds probably representing switching current directions, occur at the top of SMB-1, in turn overlain by tens of cm-thick little-disturbed stratigraphically consistent argillite.

SMB-2. The stratigraphically upper cycle locally includes a thick (<10 cm) densely packed spherule unit at the base, or isolated microkrystite-bearing carbonate bands, locally including flattened mm-scale siliceous peloids. The spherule unit is overlain by current-rippled argillite which reflects significantly lower current intensity as compared to current structures within the lower cycle SMB-1. Simonson (1992) documented splitting of the SMB at its westernmost occurrences into two parts: (1) lower argillite containing many decimeter-long and up to 25 mm-thick spherule-bearing lenses; (2) middle <7 cm thick laminated calc-argillite; (3) upper <60 cm-thick cross-layered dolomitic turbidite unit.

10.2 Late Archaean Fortescue and Hamersley Basins Impact Units

This author regards the SMB as having been emplaced either in pulses or by several closely spaced events (p. 835, op cit.). Both SMB-1 and SMB-2 can be identified in Munjina Gorge and Wittenoom Gorge – located about 40 km apart (Figs. 10.1 and 10.2). In both areas, undisturbed argillites 1–50 cm-thick separate the spherule units/lenses and current-rippled arenite/turbidite of the SMB-1 and SMB-2 units (Figs. 10.9 and 10.10). The close coupling of spherule deposition and high-energy current activity as two distinct cycles, each comprising a basal spherule unit overlain by current-disturbed turbidite, over a distance of at least 300 km, is difficult to reconcile with current-redistribution of a single original impact fallout layer. Redistribution by erosion and redeposition may be expected to be more sporadic, ensuing in places in elimination and in other places in the many-fold redeposition of spherules as well as their corrosion and breakage, a model difficult to reconcile with the consistent occurrence of two spherule-bearing units, each closely followed by strong current activity, throughout much of the Hamersley Basin.

Other essential components of SMB-1 are fragmental to irregular shaped microtektites consisting of random to crudely oriented microlites of K-feldspar and in some instances agglutinated with microkrystite spherules (Figs. 10.11 and 10.12). SMB-2 contains (1) spherule-rich units (2) mm-scale to cm-scale bands which contain isolated spherules and distinct flattened silicate-carbonate up to 3 mm long peloids of black appearance in outcrop. A complete continuum is observed between the various types of spherule-bearing units.

The microkrystite spherules display a wide range of textures, all of which contain all or some of the following features:

1. Inward-radiating fibrous/acicular K-feldspar which either form shells, shells and mantles, or completely fill the spherules (Fig. 10.12);
2. Internal voids filled with quartz, carbonate, chlorite, sericite and Fe-oxides; internal central to offset vesicles or bubbles filled mainly with quartz and carbonate. The bubbles have distinct boundaries and are different in texture from the central voids above.
3. Composite particles consisting of agglutinated microkrystite spherules and microtektites (Figs. 10.7 and 10.11)
4. Some microkrystites have both inward-radiating K-feldspar fans and randomly oriented K-feldspar needles, showing transitional characteristics with microtektites.

SEM/EDS analyses identify the distribution of K-feldspar in radiating fans in microkrystite spherules as microlites in microtektites and as micron-scale grains in the matrix. Cores may be occupied by K-feldspar, quartz, carbonate, iron oxides and accessory minerals. Inter-spherule matrices consist of microcrystalline assemblages of K-feldspar, quartz, chlorite, sericite, iron oxides and trace apatite and monazite similar to those of the host argillite and arenite. Iron oxide aggregates occur in inter-spherule position in association with microcrystalline feldspar and quartz. Flattened peloids consisting of almost pure silica, or feldspar, or silicate carbonate mix are set in matrices dominated by carbonate and chlorite. XRD analyses indicate abundance of dolomite, lesser concentration of K-feldspar and trace mica. Microkrystites display

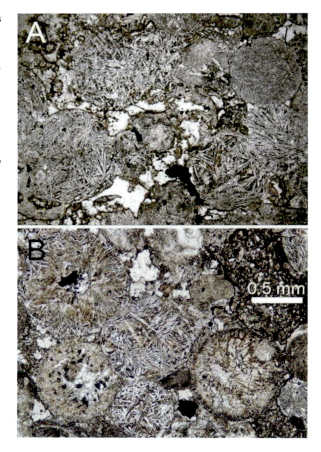

Fig. 10.11 Photomicrographs of Paraburdoo and Reivilo spherule layers in plane polarized light.
(**a**) Paraburdoo spherule layer showing tightly packed spherules with abundant K-feldspar crystallites, mix of intact and deformed spherule shapes, and poreshaped patches of sparry carbonate cement. (**b**) Reivilo spherule layer showing similar textures; scale bar applies to both images (From Hassler et al. 2011; Geological Society of America, by permission)

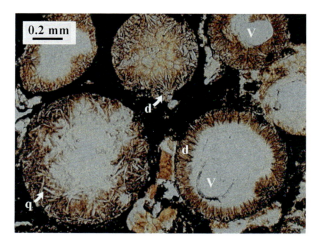

Fig. 10.12 Spherules of the ~2.56 Ga Spherule Marker Bed, showing quench (*q*) and inward-divergent (*d*) K-feldspar crystallites surrounding silicified cores and vesicles (v)

10.2 Late Archaean Fortescue and Hamersley Basins Impact Units

Table 10.3 Siderophile element abundances of spherule-bearing sediments from the ~2.57 Ga Paraburdoo spherule layer (Goderis et al. 2013) and ~2.56 Ga SMB spherule unit (Simonson et al. 1998, 2009a, b)

	Ni ppm	Co ppm	Cr ppm	Ni/Co	Ni/Cr	Ir ppb	Pt ppb	Pd ppb	Pd/Ir	Pt/Ir	Pd/Pt
Av of 12 Paraburdoo samples	404	72	1,107	5.6	0.36	247	341	81.8	0.33	1.38	0.24
Av of 9 SMB samples	36	18	100	2	0.36	0.496	5.42	1.42	2.8	10.9	0.26
BB	42.4	11.3	97	3.75	0.43	0.38	0.97	2.06	5.4	2.5	2.1
S64-1 SL	30.4	8.9	114	3.41	2.7	0.68	1.73	4.47	6.6	2.5	2.6
S77-1 SL	20.2	6.5	47	3.1	0.43	0.4	1.7	4.12	10.3	4.2	2.4
A – C1	10,500	500	2,650	21	4	455	1,010	550	1.2	2.2	0.54
B – pyrolite	1,960	105	2,625	18.5	0.75	3.2	7.1	3.9	1.2	2.2	0.55
C – mean ultramafic	2,000	110		18.2		2.4	10	8.2	3.4	4.2	0.82
D – komatiite	1,530	104		14.7	1.0	9.3	8.4	8.4	9.3	0.9	
E – MORB	200	41		4.9		0.05	2.3	0.6	12	46	0.26
F – Archaean tholeiite	<150	60–80	<100	1.8–2.5	~1.5						
G – Archaean -high-Mg	198	49	749	4	0.26	0.49		18.2	37		

Data compared to C1-chondrites (A – Chou 1978); mantle pyrolite (B – McDonough and Sun 1995); mean ultramafic (C – Chou 1978); komatiite (D – Chou 1978); mid-ocean ridge basalt (E – Chou 1978); average Archaean tholeiite (F – BVSP 1981); high-Mg basalt (G – Glikson and Hickman 1981 – mean of 20 high-Mg basalts of the Apex Basalt)

a high degree of sphericity, D_{min}/D_{max} being generally higher than 0.9. Frequency size analysis of spherules indicates maxima in the range of 500–800 µm (Fig. 10.12).

Whole rock analyses for siderophile elements (Ni, Co), Cr, Zn, and the Platinum Group Elements (PGE) reported by Simonson et al. (1998) indicate weak but significant anomalies consistent with a meteoritic derivation of the microkrystite spherules and microtektite-rich sediments. Comparative chondrite-normalized abundance profiles for the elements Ir, Ru, Pt, Pd, Au, Ni, Co, V and Cr – corresponding to progressive enrichment sequence in Pyrolite model mantle (Ringwood 1975) relative to C1-chondrite (McDonough and Sun 1995), allow the following observations (Tables 10.1, 10.2, 10.3, and 10.4):

1. Ir levels of microkrystite spherule-rich sediments (range 0.02–1.54 ppb; mean of 9 samples – 0.496 ppb) are about an order of magnitude higher than in background shale/carbonate sediments (mean of 11 samples – 0.0033 ppb) associated with the SMB.
2. Ru levels likewise tend to be higher in the microkrystite-rich sediments (range 0.07–2.61 ppb; mean 1.21 ppb) relative to background sediments (0.5 ppb).
3. Pd/Ir, Pt/Ir are distinctly lower than those of spherule-poor sediments and background sediments, namely they are closer to chondritic values (Fig. 10.13).

Table 10.4 Siderophile and PGE trace element comparisons between the mean composition of spherules from DGS4

DGS4	Ni ppm	Co ppm	Cr ppm	Ni/Co	Ni/Cr	Ir ppb	Pt ppb	Pd ppb	Pd/Ir	Pt/Ir	Pd/Pt	Ref
DGS4 Dale Gorge Member	365	11	239	44	1.53		42	2.5			0.06	1
96,357	150	7.3	54	20.5	2.8	1.13	2.37	0.64	0.57	2.1	0.27	2
96460A	422	42.5	265	9.9	1.6	13.7	28.2	7.72	0.56	2.0	0.27	2
96,466	311	36.5	272	8.5	1.1	17.9	40.8	25.1	1.4	2.0	0.61	2
DG crop	329	54.7	268	6.0	1.22	15.5	30.7	15.7	1.01	2.0	0.51	2
DG core	271	37.6	231	7.2	1.17	9.34	18.6	7.69	0.82	2.0	0.41	2
A – C1	10,500	500	2,650	21	4	455	1,010	550	1.2	2.2	0.54	
B – pyrolite	1,960	105	2,625	18.5	0.75	3.2	7.1	3.9	1.2	2.2	0.55	
C – mean ultramafic	2,000	110		18.2		2.4	10	8.2	3.4	4.2	0.82	
D – komatiite	1,530	104		14.7		1.0	9.3	8.4	8.4	9.3	0.9	
E – MORB	2,00	41		4.9		0.05	2.3	0.6	12	46	0.26	
F – Archaean tholeiite	<150	60-80	<100	1.8-2.5	~1.5							
G – Archaean -high-Mg	198	49	749	4	0.26	0.49		18.2	37			

Simonson et al. (1998) [Ref 1] and Simonson et al. (2009a, b) [Ref 2], C1-chondrites (A – Chou 1978); mantle pyrolite (B – McDonough and Sun 1995); mean ultramafic (C - Chou 1978); komatiite (D – Chou 1978); mid-ocean ridge basalt (E – Chou 1978); average Archaean tholeiite (F – BVSP 1981); high-Mg basalt (G – Glikson and Hickman 1981 – mean of 20 high-Mg basalts of the Apex Basalt)

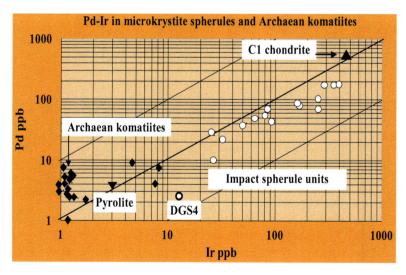

Fig. 10.13 Pd-Ir relations in Dales Gorge Shale Macroband 4 (DGS4) microkrystites compared to Barberton microkrystite spherules and Archaean komatiites. Key: *open circles* – microkrystite spherules; solid diamonds – Archaean komatiites

10.2 Late Archaean Fortescue and Hamersley Basins Impact Units

Fig. 10.14 Outcrops of the Dales Gorge Shale Macroband (DGS4) impact ejecta unit. (**a**) Cliff outcrop at Wittenoom Gorge, near Mine Pool; (**b**) DGS4 impact ejecta unit at Yampire Gorge; (**c**) DGS4 impact ejecta unit at Dales Gorge; (**d**) boulder of banded chert within DGS4 ejecta, near Mine Pool, Wittenoom Gorge; (**e**) boulder of layered chert within DGS4 ejecta, near Mine Pool, Wittenoom Gorge

However, contrary to an expectation of high siderophile element levels, the abundances of Ni (mean – 36 ppm) and Co (mean – 18 ppm) in spherule-bearing sediments are similar to those in background sediments (mean Ni – 31 ppm, mean Co – 20 ppm), whereas the mean Cr values of the spherule-rich sediments (100 ppm) is about double that of background sediments (mean – 47 ppm) (Simonson et al. 1998). Likewise, an overall overlap is observed regarding Ni/Co ratios, whereas Ni/Cr values of spherule-rich sediments are lower by about a factor of two compared to background sediments, reflecting the higher Cr values of spherule-bearing units. Whereas the absolute level of PGEs is generally low, the ratios Pd/Ir and Pt/Ir are significantly lower than those of basalts and komatiites (Tables 10.1, 10.2, 10.3), consistent with preferential condensation of the refractory PGE relative to volatile PGEs (Glikson and Allen 2004; Glikson 2007).

10.2.4 ~2.48 Ga Impact, Dales Gorge (DGS4)

LaBerge (1966) documented mm-scale spherulitic textures within the Shale-unit-4 (DGS4) Macroband of the lower Dales Gorge Member, Brockman Iron Formation (Fig. 10.14), dated as 2,479 Ga-old (Trendall et al. 1998; Thorne and Trendall 2001).

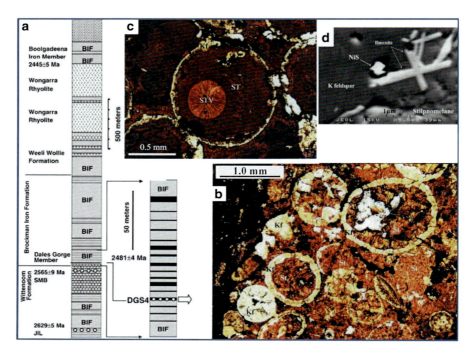

Fig. 10.15 ~2.48 Ga Dales Gorge Shale Macroband 4 (DGS4) impact ejecta, Brockman Iron Formation. (**a**) schematic columnar stratigraphy showing main units of the Hamersley Group and U–Pb zircon isotopic ages (after Trendall et al. 2004); (**b**) DGS4 Microkrystite spherules occupied by stilpnomelane (*red*), and rimmed by K-feldspar; (**c**) DGS4 Microkrystite spherules occupied by stilpnomelane (*red*), containing a central vesicle and rimmed by K-feldspar (*yellow*); (**d**) Ilmenite crystals and NiS crystal within K-feldspar

Trendall and Blockley (1970) referred to this unit as a spherule-bearing breccia. Simonson (1992) identified the spherules as microkrystites on the widespread occurrence of spherule shells consisting of inward radiating K-feldspar fans surrounding stilpnomelane-dominated interiors (Fig. 10.15) and from their morphological parameters ($D_{mean} \sim 0.7$ mm; $D_{max} < 1.8$ mm). Whereas the majority of spherules show high sphericities, elongate and dumbbell-shaped spherules as long as 2.5 mm are present. Further petrological and geochemical studies established the extraterrestrial connection of the microkrystites (Glikson and Allen 2004)

The DGS4-Macroband which hosts the spherule-bearing unit can be traced at least 30 km in the Dales Gorge-Wittenoom Gorge area (Fig. 10.1). The Macroband forms a 1–2 m-thick soft zone consisting of finely laminated siltstone and shale which form a distinct recess within cliff-forming banded iron (Fig. 10.14a). Accessible type outcrops of the DGS4 occur at Dales Gorge, Yampire Gorge and Wittenoom Gorge. The spherule-bearing unit forms a conformable stratiform zone within the siltstone/shale, and varies in thickness from about 20–30 cm at Dales Gorge to 110 cm at Wittenoom Gorge. The spherule-bearing unit forms a soft to crumbling, black-weathered, irregularly non-laminated zone, which can be easily mistaken for tachylite-rich pyroclastics. Where observed, the unit conformably overlies siltstone and shale. The top of the unit may contain cm-scale black fragments.

A distinct feature of the DGS4 is the presence of isolated angular to sub-rounded fragments and meter-scale rafts of chert and banded ferruginous chert (Fig. 10.14d) (Hassler and Simonson 2001; Glikson and Allen 2004; Glikson 2004a, b). The clasts rest at a variety of angles to the overall layering, including near-vertical orientations, resulting in flexuring of the overlying siltstone/shale. The mega-clasts appear to have been incorporated contemporaneously with the spherule-rich material and are interpreted as exotic tsunami-transported blocks ripped off submarine scarps, possibly derived from impact-triggered fault scarps. In hand specimen the black-aphanitic spherule-bearing material consists of largely obliterated fragments of stilpnomelane and isolated to tightly packed spherules marked by white K-feldspar shells. Where no K-feldspar shells are present the dark stilpnomelane-rich spherules are difficult or impossible to distinguish from the dark stilpnomelane-rich matrix. Microscopic observations reveal a wide range of spherule morphologies (Fig. 10.15) including oblate, disc-shaped and near dumbbell-like forms.

Spherules of the DGS4 Macroband display a frequency size peak in the range of 600–800 µm, although more than about 10 % of spherules are larger than 1.0 mm, including large oblate, disc-shaped and dumbbell-like spherules reaching maximum long-axis size of about 2.5 mm. The elongate spherule shapes are regarded as primary as little or no penetrative deformation has affected the sediments. K-feldspar-mantled microkrystites consist of inward-radiating K-feldspar, which may also occupy spherule interiors, otherwise occupied by stilpnomelane. Spherule interiors may contain central to offset vesicles filled with randomly oriented fans of stilpnomelane (Fig. 10.15). Secondary micro-spherules of stilpnomelane may occur within K-feldspar shells and mantles (Fig. 10.15). K-feldspar appears to display little effects of alteration or weathering. Micron-scale Nickel particles within K-feldspar include Ni-metal, Ni-oxide, Ni-sulphide and Ni-arsenide (Fig. 10.15). Where K-feldspar shells do not exist, the material may show palimpsest outlines of shell-free spherules and/or partly obliterated mm-scale to cm-scale fragmental entities, possibly representing altered glass shards.

The meteoritic impact connection of spherules included in the DGS4 Macroband of the Dales Gorge Member is demonstrated by unique mineralogical and geochemical features, including:

1. Ni-rich submicron metal, oxide, sulphide and arsenide submicron particles within K-feldspar spherule shells.
2. High Ni levels in Fe-oxide grains.
3. High Ni abundances, very high Ni/Co ratios and high Ni/Cr ratios in stilpnomelane spherule interiors.
4. Low Pd/Ir, Pt/Ir and Pd/Pt ratios in stilpnomelane spherule interiors and to a lesser extent in K-feldspar shells.
5. Inward-radiating K-feldspar fans and offset vesicles within the spherules.

SEM/EDS analyses indicate an abundance of euhedral ~5–10 μm-long ilmenite needles within K-feldspar spherule shells (Fig. 10.15). Associated with the ilmenite are rare high electron-density submicron Ni-rich particles, including near-pure Ni-metal (Ni < 84.7 %), Ni-S-As particles, Ni-Co-As-S particles and NiS particles (Fig. 10.15). Fe-oxide grains within the stilpnomelane contain an admixed silicate component and some are characterized by Ni contents up to 1.25 % NiO. Mean analyses of stilpnomelane and K-feldspar from four laser-ICPMS sections across microkrystites indicate meteoritic geochemical signatures, including high siderophile elements (Ni, Co, Cr) and high PGE abundances. Stilpnomelane spherule cores display Ni levels (~327–436 ppm, mean ~365 ppm), higher than basaltic high Cr basalts (47–387 ppm; mean ~240 ppm), and very low Co levels (mean ~11 ppm), with ensuing very high Ni/Co (mean ~44) and high Ni/Cr (~1–10; mean ~4) ratios as compared to basalts.

Whole rock PGE measurements by solution ICPMS yield high abundances of Ir (range ~1–18 ppb), Pt (range ~2–42 ppb) and distinctly low Pd (range ~0.64–25.1 ppb), reflected by low Pd/Ir ratios (0.56–1.4) as compared to terrestrial mafic rocks (range 3.4–37) (Table 10.4). The relative depletion in volatile PGEs is consistent with PGE patterns in microkrystite spherules from other impact ejecta units (Glikson and Allen 2004; Simonson et al. 2009a, b) and is contrasted with the almost invariable enrichment of volatile PGEs (Pd, Au) in terrestrial materials, with the notable exception of refractory harzburgites (Chou 1978). In agreement with the above the C1-chondrite-normalized plot for the siderophile elements, following a pyrolite/chondrite depletion sequence (Ir < Pt < Pd < Au < Ni < Co < V < Cr < Sc), indicates significant departures from pyrolite ratios, including the high Ni/Co and low Pd/Ir and Pt/Ir noted above. Au abundances are not included in this discussion due to the high secondary mobility of this element under hydrous conditions.

The DS4 impact fallout unit includes larger microkrystite spherules than observed in any other Pilbara unit, including spherule diameters in the range of <1.4 mm and long axes of oblate and dumbbell-shaped spherules up to <2.2 mm. In accord with thermodynamic estimates by O'Keefe and Ahrens (1982) and Melosh and Vickery (1991) spherule condensates on this scale represent projectile diameters in the order of 15–25 km.

The likely equivalent of DGS4 were encountered in a drill hole ~30 km south of Griquastad, western Transvaal Basin, where a ~1 cm-thick spherule layer is located below an ~80 cm-thick breccia unit within a 2 m-thick shale unit near the base of the Kuruman Iron Formation, about 37 m above Gamahuann carbonates.

10.2.5 Estimates of Asteroid Size and Compositions

Mass balance calculations based on the high PGE contents of microkrystite spherules (Tables 10.1 to 10.4; Fig. 10.16) support asteroid sizes on the order of tens of km (Byerly and Lowe 1994; Shukolyukov et al. 2000; Kyte et al. 2003, 2004). Size estimates presented below include the following assumptions: (1) derivation of the

10.2 Late Archaean Fortescue and Hamersley Basins Impact Units

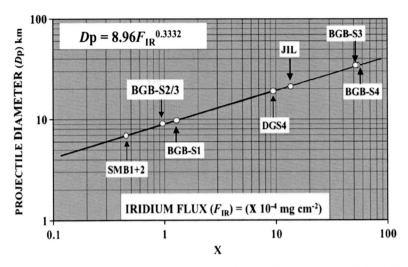

Fig. 10.16 Correlation between Ir flux (in units of 10^{-4} mg cm^{-2}) and diameter of chondritic projectile (Dp), based on mass-balance calculations assuming mean unit thickness, mean Ir concentration, and global distribution of ejecta. Unit symbols: $\sqrt{[Vp/(4/3)\pi]}$; $Vp = Mp/dp$; $Mp = FG/Cp$; FG Earth $= AE/dS \times TS \times S$Earth (Earth's surface area), where AE is measured element (E) abundance in ejecta unit (in ppb); dS is mean density of ejected materials (mg/cm^3) (assumed as 2.65 g/cm^3); CE = mass of element E (in mg/cm^3); (CE = AE × DS); TS = mean stratigraphic thickness of spherule unit (in mm); FE = local mean flux of element E (in mg per cm^2 surface; FE = CE × TS); FG = inferred global flux (in mg/cm^2) of element E; Cp = assumed concentration of element E in projectile (ppb) (C1 chondrites have 450 ppb Ir); Mp = mass of projectile; dp = assumed density of projectile (C1 chondrites' density = 3.0 g/cm^3) (Elsevier, by permission)

bulk of the PGE from the projectile; (2) chondritic projectile with 500 ppb Ir and 1,010 ppb Pt; (3) global fallout unit thickness 50 % of the minimum thickness observed in outcrop or drill holes. Particularly the last assumption, although conservative, is fraught with uncertainty – rendering such estimates highly tentative.

The mass balance equation is formulated as:

$$Rp = \sqrt{[Vp/(4/3)\pi]};$$
$$Vp = Mp/dp; Mp = FG/Cp; FG_{Earth} = AE/dS \times TS \times S_{Earth} \text{ (Earth's surface area)}$$

where AE is measured element (E) abundance in ejecta unit (in ppb); dS is mean density of ejected materials (mg/cm^3) (assumed as 2.55 g/cm^3); CE = mass of element E (in mg/cm^3); (CE = AE × DS); TS = mean stratigraphic thickness of spherule unit (in mm); FE = local mean flux of element E (in mg per cm^2 surface; FE = CE × TS); FG = inferred global flux (in mg/cm^2) of element E; Cp = assumed concentration of element E in projectile (ppb) (for C1 chondrites with 450 ppb Ir); Mp = mass of projectile; dp = assumed density of projectile (C1 chondrites' density = 3.0 g/cm3).

Based on these calculations estimates of projectile diameters are given in Fig. 10.16, which should however be regarded as tentative order of magnitude figures. Mass balance calculations derived from PGE and $^{53}Cr/^{52}Cr$ isotopes a carbonaceous chondrite composition was suggested for the Barberton 3.225 Ga S3 and S4 spherules (Kyte et al. 2003), a carbonaceous chondrite suggested for the 65 Ma K-T boundary impact (Shukolyukov and Lugmair 1998) and ordinary chondrite type suggested for the Late Eocene impacts (Kyte et al. 2004).

10.3 Impact Fallout Units and Banded Iron Formations

A temporal association observed between Archaean to earliest Proterozoic asteroid impact ejecta/fallout units and overlying banded iron formations (Glikson 2006; Glikson and Vickers 2007) may hint that, in some instances, these impacts were closely followed by significant transformation in the nature of source terrains of the sediments. Recorded relations between impact spherule layers and banded iron-formation (BIF), Jaspilite and ferruginous shale include the 3,470.1 ± 1.9 Ma spherule units of the Antarctic Chert Member, Mount Ada Basalt (Glikson et al. 2004) (Fig. 8.14). The 3,243 ± 4 Ma impact unit (S3) in the Barberton Greenstone Belt (BGB) is overlain by iron-rich sediments (Ulundi Formation). The 3,258 ± 3 Ma impact spherule unit (S2) of the BGB is overlain by BIF, Jaspilite and ferruginous shale. Two BIF and ferruginous shale units occur at the base of the Gorge Creek Group (Nimingarra Iron Formation and Paddy Market Formation), central Pilbara Craton, which overlie 3,235 ± 3 Ma felsic volcanics associated with an unconformity and olistostrome correlated with the BGB-S3 impact unit (Sect. 7.3).

The Jeerinah Impact Layer (JIL) (Simonson et al. 2000a, b; Simonson and Glass 2004; Glikson 2004a, b; Hassler et al. 2005; Rasmussen and Koeberl 2004; Rasmussen et al. 2005) overlies an argillite-dominated unit (Jeerinah Formation, 2,684 ± 6 Ma (Trendall et al. 2004) and lies directly below a thin volcanic tuff (2,629 ± 5 Ma), overlain by the Marra Mamba Iron Formation (MMIF) (2,597 ± 5 Ma) comprising banded iron formation (BIF). The Spherule Marker Bed (SMB) (Simonson 1992; Simonson and Hassler 1997; Simonson and Glass 2004), which includes two impact cycles (Glikson 2004a, b) is located at the top of a carbonate/calcareous siltstone-dominated sequence (Bee Gorge Member, Wittenoom Formation, 2,565 ± 9 Ma (Trendall et al. 2004) and below a carbonate-poor siltstone–chert–BIF sequence (Mount Sylvia Formation, Bruno's Band (BIF), Mount McRae Shale, 2,504 ± 5 Ma (Rasmussen et al. 2005). A unit of impact spherules (microkrystite) correlated with JIL is observed within a more than 100 m-thick fragmental-intraclast breccia pile in drill cores near Roy Hill, Eastern Hamersley Basin. The sequence represents significant thickening of the impact/tsunami unit relative to the JIL type section at Hesta, central Pilbara, as well as relative to the 20–30 m-thick ~2.63 Ga Carawine Dolomite spherule bearing mega-breccia (SBMB) Oakover River to the East. The location of the JIL and associated breccia under supergene iron ore within the Marra mamba Iron Formation at Roy Hill renders the impact products of economic interest. The scarcity, with few exceptions

(Rasmussen and Koeberl 2004), of shocked quartz grains in the ejecta, suggests impacts occurred in SIMA regions of the late Archaean Earth (Simonson et al. 1998; Glikson 2005).

No significant thicknesses of iron-rich sediments are known to overlie carbonate-hosted impact ejecta units, including the 2,561 ± 8 Ma Spherule Marker Bed (Bee Gorge Member) of the Wittenoom Formation, Monteville Formation and Reivilo Formation (Griqualand West Basin, Transvaal Group). However, the carbonate-shale sequence overlying the SMB is noticeably more ferruginous than underlying carbonates (Glikson and Vickers 2007). Impact spherules associated with the 4th Shale Macroband of the Dales Gorge Iron Member (DGS4) of the Brockman iron Formation. The ~2.48 Ga-old Dales Gorge Member (DGS4) of the Brockman Iron Formation is underlain by a ~0.5 m-thick rip-up clast breccia located at the top of the ~2.50 Ga Mt McRae Shale, and is interpreted as a tsunami deposit.

Barring a possible presence of undocumented hiatuses between the impact layers and directly overlying units, such as in the case of the JIL-MMIF relations, and within the accuracy limits of U–Pb zircon age data, it may follows the JIL and SMB mega-impacts were succeeded by enhanced supply of ferruginous and clastic materials, possibly reflecting enrichment of sea water in soluble ferrous iron derived from impact-triggered mafic volcanic and hydrothermal activity. A possible explanation of such links is provided by mafic volcanism triggered by large asteroid impacts, enriching the oceans in soluble FeO under Archaean atmospheric conditions, followed by seasonal microbial and/or photolytic oxidation to ferric oxide and precipitation of Fe2O3 and silica.

10.4 Inter-continental Correlation of Impact Units

As well as furnishing regional stratigraphic markers, impact fallout units allow intercontinental time-event correlations, including (Fig. 10.2):

1. A correlation between the chert/arenite impact fallout unit of the 3,470.1 ± 1.9 Ma Antarctic Chert Member of the Mount Ada Basalt and the S1 unit of similar age in the Onverwacht Group, Barberton Mountain Land, Transvaal (Byerly et al. 2002) (Figs. 7.10 and 8.3).
2. A correlation between the ~3,243 ± 3 Ma Sulphur Springs Group – Gorge Creek Group break and the 3,237 ± 6–3,225 ± 3 Ma Onverwacht Group – Fig Tree Group break, Barberton Mountain Land, Transvaal (Lowe et al. 1989, 2003; Glikson and Vickers 2006).
3. A possible correlation between the (2,629 ± 5 Ma) Ga Jeerinah Impact Layer (JIL) at the top Jeerinah Formation and the impact fallout unit in the Monteville Formation (<2,650 ± 8 Ma), Western Transvaal (Simonson et al. 1999, 2009b) (Fig. 10.2).
4. A possible correlation between the 2,481 ± 4 Ma Dales Gorge Shale-4 Macroband (DGS4) and a 10 cm-thick + spherule + breccia unit at the lower part of the Kuruman Iron Formation (<2,460 ± 5 Ma, near Griqualand, western Transvaal Basin (Simonson et al. 2009b).

Chapter 11
The Early Atmosphere and Archaean Life

Abstract The application of isotopic tracers to paleo-climate investigations, including oxygen ($\delta^{18}O$), sulphur ($\delta^{33}S$) and carbon ($\delta^{13}C$), integrated with Sedimentological and proxies studies, allows vital insights into the composition of early atmosphere–ocean-biosphere system, suggesting low atmospheric oxygen, high levels of greenhouse gases (CO_2, CO, CH_4 and likely H_2S), oceanic anoxia and high acidity, limiting habitats to single-cell methanogenic and photosynthesizing autotrophs. Increases in atmospheric oxygen have been related to proliferation of phytoplankton in the oceans, likely about ~2.4 Ga (billion years-ago) and 0.7–0.6 Ga. The oldest recorded indirect traces of biogenic activity are provided by dolomite and banded iron sediments (BIF) from ~3.85 Ga-old Akilia and 3.71–3.70 Ga Isua greenstone belt, southwest Greenland, where metamorphosed banded ironstones and dolomite seawater-like REE and Y signatures (Bolhar et al, Earth Planet Sci Lett 222:43–60, 2004; Friend et al. 2007) were shown to be consistent with those of sea water (Nutman et al, Precamb Res 183:725–737, 2010). Oldest possible micro-fossils occur in ~3.49 Ga black chert in the central Pilbara Craton (Glikson 2010; Duck et al. Geochim Cosmochim Acta 70:1457–1470, 2008; Golding et al, Earliest seafloor hydrothermal systems on earth: comparison with modern analogues. In: Golding S, Glikson MV (Eds.) Earliest life on earth: habitats, environments and methods of detection. Springer, Dordrecht, pp 1–15, 2011) and in 3.465 Ga brecciated chert (Schopf et al, Precamb Res 158:141–155, 2007). Possible stromatolites occur in ~3.49 and ~3.42 carbonates. The evidence suggests life may have developed around fumaroles in the ancient oceans as soon as they formed.

Keywords Early atmosphere • Carbon dioxide • Methane • Oxygen • Early life • Stromatolite • Microfossils

11.1 Archaean Carbon-Oxygen-Sulphur Cycles and the Early Atmosphere

During early stages of terrestrial evolution low solar luminosity (The Faint Young Sun, 27 % lower luminosity than at present) (Sagan and Mullen 1972), representing a progressive increase in fusion of hydrogen to helium, was compensated by high greenhouse gas (GHG) levels (Fig. 11.1), allowing surface temperature to remain above freezing. Alternative hypotheses were proposed by Longdoz and Francois (1997) in terms of albedo and seasonal variations on the early Earth and by Rosing et al. (2010) in terms of a high ocean/continent surface area ratio in the Archaean, leading to lower albedo and absorption of infrared by open water. Temporal fluctuations in atmospheric GHG levels constituted a major driver of alternating glacial and greenhouse states (Kasting and Ono 2006). Knauth and Lowe (2003) and Knauth (2005) measured low $\delta^{18}O$ values in ~3.5–3.2 Ga cherts of the Onverwacht Group, Barberton Greenstone Belt (BGB), Kaapvaal Craton, suggesting extremely high ocean temperatures in the range of 55–85 °C. The maximum $\delta^{18}O$ value in Barberton chert (+22‰) is lower than the minimum values (+23‰) in Phanerozoic sedimentary cherts, precluding late diagenesis as the explanation of the overall low $\delta^{18}O$ values. Regional metamorphic, hydrothermal, or long-term resetting of original $\delta^{18}O$ values is also precluded by preservation of $\delta^{18}O$ across different

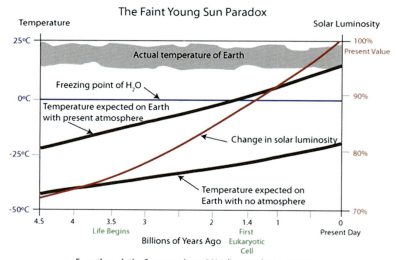

Fig. 11.1 The faint young sun paradox according to Sagan and Mullen (1972), suggesting compensation of the lower solar luminosity by high atmospheric greenhouse gas levels at early stages of terrestrial evolution (Reproduced from The Habitable Planet course produced by Harvard Smithsonian Center for Astrophysics with permission from the Annenberg Foundation, www.learner.org. Courtesy Michele McLeod)

11.1 Archaean Carbon-Oxygen-Sulphur Cycles and the Early Atmosphere

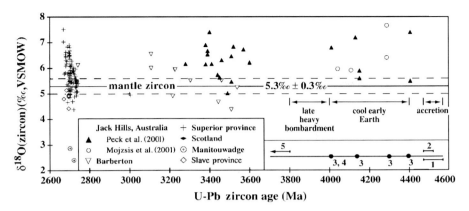

Fig. 11.2 Crystallization age (*U-Pb*) and $\delta^{18}O$ for Archaean magmatic zircons, the latter showing values mainly similar to present mantle zircon values and little variation throughout the Archaean. Some zircon $\delta^{18}O$ values are as high as 7.5‰, regarded the result of melting of protoliths altered by interaction with low temperature water near the surface (From Valley et al. 2002; Geological Society of America, with permission)

metamorphic grades. According to Knauth (2005) high-temperature conditions extended beyond submarine fumaroles and the Archaean oceans were characterized by high salinities 1.5–2.0 times the modern level. In this interpretation ensuing evaporite deposits were removed by subduction, allowing lower salinities. However, well preserved Archaean sedimentary sequences contain little evidence of evaporite deposits. The low-oxygen levels of the Archaean atmosphere and hydrosphere limited marine life to extremophile cyanobacteria. Microbial methanogenesis involves reactions of CO_2 with H_2 and in organic molecules produced from fermentation of photo synthetically produced organic matter. Photolysis of methane may have created a thin atmospheric organic haze.

Studies of oxygen isotopes of Hadean zircons indicate little difference between their maximum $\delta^{18}O$ values of ~4.4–4.2 Ga zircons and those of younger ~3.6–3.4 Ga zircons (Fig. 11.2), which suggests presence of relatively low temperature water near or at the surface (Valley et al. 2002, 2005). An overall increase with time in $\delta^{18}O$ shown by terrestrial sediments (Valley 2008) (Fig. 11.3) reflects long term cooling of the hydrosphere, consistent with an overall but intermittent temporal decline in atmospheric CO_2 shown by plant leaf pore (stomata) proxy studies (Berner 2004, 2006; Beerling and Berner 2005; Royer et al. 2001, 2004, 2007). The long-term decline may have been associated with increased rates of weathering-sequestration of CO_2 related to erosion of rising orogenic belts, including the Caledonian, Hercynian, Alpine, Himalayan and Andean mountain chains (Ruddiman 1997, 2003). Such a trend is consistent with suggested increase in the role plate tectonics through time (Glikson 1980), which led to an increase in sequestration of CO_2 by weathering of uplifted orogenic belts and subduction of CO_2-rich carbonate and carbonaceous shale. Central to studies of early atmospheres is the level of oxygen and its relation to photosynthesis. Sulphur isotopic analyses record mass-independent fractionation of sulphur

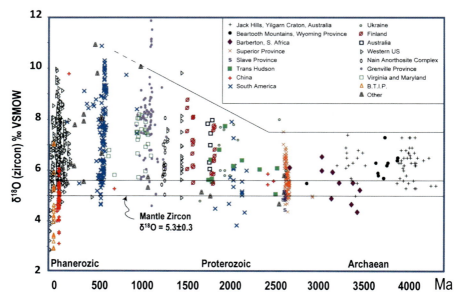

Fig. 11.3 $\delta^{18}O$ ratio of igneous zircons from 4.4 Ga to recent, displaying an increase in abundance of low-temperature effects with time from approximately ~2.3 Ga (Courtesy John Valley)

isotopes ($d^{33}S$) (MIF-S) in sediments older than ~2.45 Ga, widely interpreted in terms of UV-triggered reactions under oxygen-poor ozone-depleted atmosphere and stratosphere (Farquhar et al. 2000, 2007) (Fig. 11.4). From about ~2.45–2.32 Ga – a period dominated by deposition of banded iron formations (BIF), $d^{33}S$ values signify development of an ozone layer shielding the surface from UV radiation (Farquhar et al. 2000; Kump 2009).

Evidence for the Archaean carbon cycle is contained in carbonaceous shale whose low $\delta^{13}C$ indices are suggestive of biological activity. The carbon isotopic compositions of Archaean black shale, chert and BIF provide vital clues to the proliferation of autotrophs in the shallow and deep marine environment. Peak biogenic productive periods about 2.7–2.6 Ga are represented by low $\delta^{13}C$ of chert and black shale intercalated with banded iron formations in the Superior Province, Canada (Goodwin et al. 1976), and in the Hamersley Basin, Western Australia (Eigenbrode and Freeman 2006) (Fig. 11.5). This peak, which coincides with intense volcanic activity in greenstone belts world-wide, suggests enhanced biological activity related to volcanic emanations and enriched nutrient supply. Biological processes include oxygen capture by iron-oxidizing microbes, microbial methanogenesis producing atmospheric CH_4, microbial sulphur metabolism producing H_2S, ammonia-releasing microbes, oxygen-releasing photosynthesizing colonial prokaryotes (stromatolites) (Figs. 11.6 to 11.10) and algae, culminating in production of O_2-rich atmosphere and the O_3 ozone layer. Earliest manifestations of biological activity

11.1 Archaean Carbon-Oxygen-Sulphur Cycles and the Early Atmosphere

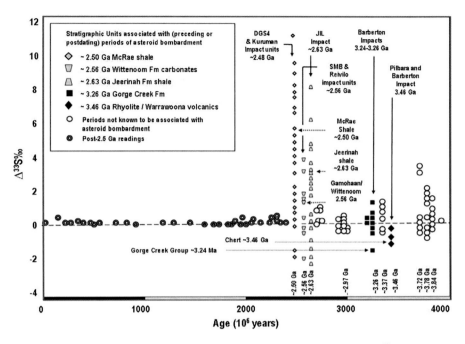

Fig. 11.4 Plots of mass independent fractionation values for Sulphur isotopes ($\delta^{33}S$ - MIF-S) vs Age. The high $\delta^{33}S$ values for pre-2.45 Ga sulphur up to about $\delta^{33}S = 11$ is interpreted in terms UV-induced isotopic fractionation, allowed by a lack of an ozone layer. Periods of major asteroid impacts during which ozone may have been destroyed are indicated (From Glikson 2010. Elsevier, by permission)

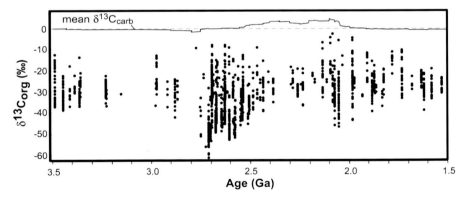

Fig. 11.5 A compilation of published kerogen and total organic carbon $\delta^{13}C$ values (δ_{org}) for all sedimentary rock types. Note high accumulations of organic carbon in the late Archaean at ~2.7–2.6 Ga (From Eigenbrode and Freeman 2006; PNAS, by permission)

Fig. 11.6 Stromatolite-like forms in intercalated barite and chert, ~3.49 Ga Dresser Formation, central Pilbara. (**a**) Cross layering view; (**b**) parallel-layering view; (**c**) an individual stromatolite-like structure; (**d**) a polished stromatolite-like form. Courtesy Bill Schopf

Fig. 11.7 Outcrop of ~3.43 Ga stromatolite-like structures at the Trendall locality, Shaw River, central Pilbara Craton

11.1 Archaean Carbon-Oxygen-Sulphur Cycles and the Early Atmosphere 165

Fig. 11.8 Outcrops of ~3.43 Ga stromatolite-like structures, Strelley Pool Formation, central and eastern Pilbara Craton; (**a**) Eastern Pilbara (courtesy Reg Morrison); (**b**) Trendall locality, Shaw River; (**c**) Eastern Pilbara; (**d** and **e**) Central Pilbara, south of Carlindi batholith; (**f**) Trendall locality, Shaw River

may be represented by banded iron formations, widely held to represent ferrous to ferric iron oxidation by microbial reactions (Cloud 1968, 1973; Morris 1993; Konhausser et al. 2002; Glikson 2006).

According to Kopp et al. (2005) photosynthetic oxygen release from cyanobacteria and aerobic eukaryotes affected oxidation of atmospheric methane, triggering planetary-scale glaciation at least as early as 2.78 Ga and perhaps as long ago as 3.7 Ga. According to Ohmoto et al. (2006) MIF-S values and oxygen levels fluctuated through time and place, possibly representing local volcanic activity. The role of asteroid impacts in affecting the ozone layer, UV radiation and thus regional MIF-S values has been discussed by Glikson (2010). Some of the earliest manifestations of biological activity may be represented by banded iron formations (BIF) from the ~3.8 Ga Isua supracrustal belt (southwest Greenland) (Fig. 2.7). Banded iron formations are

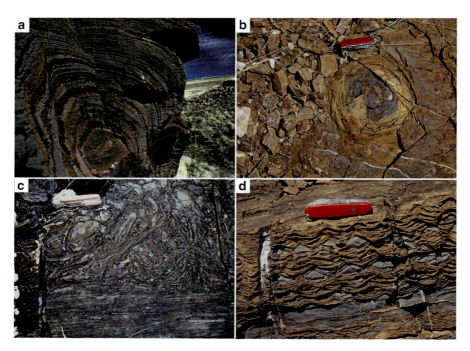

Fig. 11.9 ~2.73 Ga Stromatolites, Tumbiana Formation, Meentheena, Fortescue Basin, eastern Pilbara: (**a**) Large stromatolite near Meentheena Station, Nullagine River; (**b**) bedding plane parallel view of Meentheena stromatolites; (**c**) storm-broken rip-up stromatolite bed; (**d**) rippling of current-worked carbonate layers interspersed with Meentheena stromatolites

commonly intercalated with volcanic tuff and carbonaceous shale whose low d^{13}C indices are suggestive of biological activity. The carbon isotopic compositions of Archaean black shale, chert and BIF provide vital clues to the proliferation of autotrophs in the shallow and deep marine environment. Peak biogenic productive periods about 2.7–2.6 Ga are represented by low d^{13}C of chert and black shale intercalated with banded iron formations in the Superior Province, Canada (Goodwin et al. 1976) and in the Hamersley Basin, Western Australia (Eigenbrode and Freeman 2006). This peak, which coincides with intense volcanic activity in greenstone belts world-wide, suggests enhanced biological activity related to volcanic emanations and enriched nutrient supply. Biological processes include oxygen capture by iron-oxidizing microbes, microbial methanogenesis producing atmospheric CH$_4$, microbial sulphur metabolism producing H$_2$S, ammonia-releasing microbes, oxygen-releasing photosynthesizing colonial prokaryotes (stromatolites) and algae, culminating in production of O$_2$-rich atmosphere and the O$_3$ ozone layer. Earliest manifestations of biological forcing may be represented by banded iron formations, widely held to represent ferrous to ferric iron oxidation by microbial reactions (Cloud 1968, 1973; Morris 1993; Konhausser et al. 2002; Glikson 2006).

11.1 Archaean Carbon-Oxygen-Sulphur Cycles and the Early Atmosphere

Fig. 11.10 A giant ~40 m-large ~2.63 Ga stromatolite within the Carawine Dolomite at Carawine Pool, East Pilbara Craton; arrows point to the rim (*R*) and core (*C*) of the stromatolite; *Inset* – stromatolite within Carawine Dolomite, Woodie-Woodie area, Oakover River, eastern Pilbara

The origin of BIFs has been interpreted in terms of transportation of ferrous iron in ocean water under oxygen-poor atmospheric and hydrospheric conditions of the Archaean (Cloud 1968; Morris 1993). Oxidation of ferrous to ferric iron could occur through chemotrophic or phototrophic bacterial processes (Konhausser et al. 2002) and/or by UV-triggered photo-chemical reactions. The near-disappearance of banded iron formations (BIF) about ~2.4 Ga, with transient reappearance about 1.85 Ga and in the Vendian (650–543 Ma), likely reflect increase in oxidation, where ferrous iron became unstable in water and the deposition of BIF was replaced by that of detrital hematite and goethite. Archaean impact ejecta units in the Pilbara and Kaapvaal Cratons are commonly overlain by ferruginous shale and BIF (Glikson 2006; Glikson and Vickers 2007), hinting at potential relations between Archaean impact clusters, impact-injected sulphate, consequent ozone depletion, enhanced UV radiation and formation of BIFs (Glikson 2010), possibly by photolysis. Some of the oldest possible micro-fossils occur in black chert of the Dresser Formation (Duck et al. 2008; Golding et al. 2011) and in brecciated chert of the 3465 Ma Apex Basalt, Warrawoona Group, Pilbara Craton (Schopf et al. 2007). The paleo-environment, carbonaceous composition, mode of preservation, and morphology of these microbe-like filaments, backed by new evidence of their cellular structure provided by two- and three-dimensional

Raman imagery, support a biogenic interpretation. Evidence for hydrothermal and methanogenic microbial activity (Schopf and Packer 1987; Schopf et al. 2007; Hofmann et al. 1999; Duck et al. 2008; Golding et al. 2011) and intermittent appearance of shallow water stromatolites in ~3.49 Ga (Dunlop and Buick 1981) and ~3.42 Ga sediments (Alwood et al. 2007) testify to a diverse microbial habitat. This included heliotropic and by implication photosynthesizing stromatolites affecting release of oxygen. Problems in identifying early Archaean stromatolites were expressed by Lowe (1994) and by Brazier et al. (2002). Whereas the early stromatolites may represent Prokaryotes, Eukaryotes possibly appeared about ~2.1–1.6 Ga (Knoll et al. 2006), or earlier (Sugitania et al. 2009) and exist at present at Shark Bay, Western Australia. The occurrence of glacial stages during the Precambrian (pre-0.54 Ga) despite high atmospheric greenhouse gas levels is accounted for by Faint Early Sun conditions, examples include:

(A) Upper Archaean glaciation represented by the 5,000 m-thick Mozaan Group of the 2,837 ± 5 Ma Pongola Supergroup (Strik et al. 2007), which includes a sequence of diamictite containing striated and faceted clasts and ice-rafted debris (Young et al. 1998), representing the oldest glaciation recorded to date.
(B) An early Proterozoic Huronian glaciation (~2.4–2.2 Ga) recorded by outcrops in the North American Great Lakes district, Pilbara Western Australia and the Transvaal (Kopp et al. 2005).

The glaciation, cooling of the oceans and enrichment of oxygen in cold water led to enhanced photosynthesis by phytoplankton and thereby further enrichment of oxygen. Likewise Kasting and Ono (2006) invoke biological activity as driver of the ~2.4 Ga glaciation, including a photosynthetic rise in O_2 and concomitant decrease in CH_4.

Glacial deposits of the Cryogenian "Snowball Earth" (750–635 Ma) observed in Namibia, South Australia, Oman and Svalbard, correspond to the period of fragmentation of the long-lived Rodinia supercontinent (~1.1–0.75 Ga) (Hofmann et al. 1998). Palaeomagnetic evidence suggests that the ice sheets reached sea level close to the equator during at least two glacial episodes. Some glacial units include sedimentary iron formations, underpinning potential relations between BIF and glaciations. According to Kirschvink (1992) the runaway albedo feedback exerted by the ice sheets resulted in a near-global ocean ice cover whereas continental ice covers remained thin due to retardation of the hydrological cycle. In this model the appearance of banded iron formations may represent sub-glacial anoxia and thereby enrichment of sea water in ferrous iron. The termination of glaciation is marked by carbonates of the so-called cap carbonate (Halverson et al. 2005).

Hofmann et al. (1998) report Negative carbon isotope anomalies in carbonate rocks which cap the Neoproterozoic glacial deposits in Namibia, indicating accumulation of organic matter following glacial collapse in connection with volcanic outgassing, raising atmospheric CO_2 to some 350 times the modern level. The late Proterozoic thus represents a transition from oxygen-poor composition of early atmospheres dominated by reduced carbon species such as methane, and oceans dominated by reducing microbial processes by chemo-bacteria. Following the

11.1 Archaean Carbon-Oxygen-Sulphur Cycles and the Early Atmosphere

Cryogenian ice age ~750–635 Ma (Hofmann et al. 1998; Hofmann and Schrag 2000) the rise in oxygen during ~635–542 Ma and particularly following 580 Ma allowed oxygen-binding proteins and emergence of the multicellular Ediacara fauna in an oxygenated "Canfield Ocean". According to Canfield et al. (2007) oxygen levels constituted the critical factor allowing multicellular animals to emerge in the late-Neoproterozoic, as evidenced from the oxidation state of iron before and after the Cryogenian glaciation. A prolonged stable oxygenated environment may have permitted the emergence of bilateral motile animals some 25 million years following glacial termination, later followed by the onset of the Cambrian explosion of life (Gould 1990) from ~542 Ma and development of a rich variety of organisms.

Studies of nature of the early terrestrial atmosphere–biosphere system make essential use of sulphur, carbon and oxygen stable isotopes (Holland 1984, 1994; 2005; Pavlov and Kasting 2002; Kasting and Ono 2006). The geochemical behaviour of multiple sulphur isotopes is a key proxy for long-term changes in atmospheric chemistry (Mojzsis 2007; Thiemens 1999). The identification of mass-independent fractionation of sulphur isotopes (MIF-S) in pre-2.45 Ga sediments has been correlated with ultraviolet radiation effects on the $\delta^{33}S$ values, with implications for an ozone and oxygen-poor Archaean atmosphere (Farquhar et al. 2000, 2007), whereas other authors have suggested heterogeneous Archaean oxygen levels (Ohmoto et al. 2006). Development of photosynthesis, and thereby limited release of oxygen as early as about 3.4 Ga, is suggested by identification of heliotropic stromatolite reefs in the Pilbara Craton (Allwood et al. 2006a, b). The abrupt disappearance of positive MIF-S anomalies at ~2.45 Ga poses a problem, as atmospheric enrichment in oxygen due to progressive photosynthesis could, perhaps, be expected to result in a gradual rather than an abrupt decline in MIF-S signatures. MIF-S ($\delta^{33}S$) anomalies (Fig. 11.4) overlap mid-Archaean impact periods (~3.26–3.24 Ga) and Late Archaean impact periods (~2.63, ~2.56, ~2.48 Ga) (Lowe et al. 1989, 2003; Simonson and Hassler 1997; Simonson et al. 2000a, b; Simonson and Glass 2004; Glikson 2001, 2004a, Glikson 2005; Glikson 2006; Glikson and Allen 2004; Glikson et al. 2004; Glikson and Vickers 2006, 2007), though no specific age correlations are observed. Estimates of projectile diameters derived from mass balance calculations of iridium levels, $^{53}Cr/^{52}Cr$ anomalies and size-frequency distribution of fallout impact spherules (microkrystites) (Melosh and Vickery 1991), suggest projectiles of the ~3.26–3.24, ~2.63, ~2.56 and ~2.48 Ga impact events reached several tens of kilometre in diameter (Byerly and Lowe 1994; Shukolyukov et al. 2000; Kyte et al. 2003; Glikson and Allen 2004; Glikson 2005, 2013). Impacts on this scale would have led to major atmospheric effects, including large scale ejection of carbon and sulphur-bearing materials, effects on the ozone layer and isotopic fractionation of sulphur. Archaean impact ejecta units in the Pilbara and Kaapvaal Cratons are almost invariably overlain by ferruginous shale and banded iron formations (BIF) (Glikson 2006; Glikson and Vickers 2007). The origin of BIFs is interpreted, alternatively, in terms of oxidation of ferrous to ferric iron under oxygen-poor atmospheric and hydrospheric conditions (Cloud 1968; Morris 1993), direct chemolithotropic or photo-ferrotropic oxidation of ferrous iron (Konhausser et al. 2002), and UV-triggered photo-chemical reactions (Cairns-Smith 1978). The relations between sulphur, oxygen and carbon isotopes, atmospheric oxygen levels, photosynthesis, banded iron formations and glaciations with implications for evolution of the early atmosphere remains the subject of current investigations.

11.2 Archaean Life

> If the theory of evolution be true, it is indisputable that before the lowest Cambrian stratum was deposited, long periods elapsed … and that during these vast periods, the world swarmed with living creatures. Charles Darwin, The Origin of Species.

Inherent in the search for biological signatures in the Archaean geological record are fundamental questions associated with the origin of life (Darwin 1859; Cloud 1968; Davies 1999; Schopf 2001), conceived according to the Miller–Urey and other experiments to have arisen by racemic abiogenic synthesis of inorganic precursors to organic amino acids, the building blocks of life, triggered by lightning and radiation affecting a "primordial soup" rich in organic matter. Alexander Oparin (1924) regarded atmospheric oxygen as a constraint on synthesis of biomolecules, restricting original synthesis of biomolecules to the pre-oxygenation era. Metabolism may have occurred prior to replication, or the other way around, leading to a view of life as an extreme expression of kinetic control and the emergence of metabolic pathways manifesting replicative chemistry (Pross 2004).

According to Davies (1999) the probability of the primordial DNA/RNA biomolecules forming by accident is about $1-10^{22}$, rendering it equally or more likely the natural intelligence underlying the characteristics of these molecules and their more complicated successors resides in undecoded laws of complexity. According to Russell et al. (2010) viruses – forming parasitic entities on life forms – acted as mobile RNA worlds injecting genetic elements into proto-cells capable of replicating themselves around submarine alkaline hydrothermal vents. There chemical reactions created biosynthetic pathways leading to emergence of sparse metabolic network and the assembly of pre-genetic information by primordial cells. In this concept, life and evolution of prokaryotes constituted a deterministic process governed by bio-energetic principles likely to apply on planets throughout the universe, which means life is everywhere. However, exobiogenic theories which invoke introduction of biomolecules from other planets merely relegate the question of the origin of life further in time and space. Thus, theories suggesting seeding of life on Earth by comets are to date lacking in evidence, since while comets and meteorites may contain amino acids, the difference between amino acids and RNA or DNA is as vast as the difference between atoms of iron and supercomputers.

Much emphasis has been placed on likely analogies between extremophile microbial communities, such as around submarine hydrothermal vents ("black chimneys") and original life forms (Martin et al. 2008). These authors compare the chemistry of the H2–CO2 redox couple in hydrothermal systems and core energy metabolic biochemical reactions of modern prokaryotic autotrophs. The high greenhouse gas

11.2 Archaean Life

levels (CO_2, CO, CH_4) of the early atmosphere allowing the presence of water at the Earth surface despite low solar luminosity (Fig. 11.1) implies high partial CO_2 pressure and low pH of the Archaean oceans. Dissociation of H_2O associated with microbial Fe^{+2} to Fe^{+3} transition under these conditions would have released hydrogen (Russell and Hall 2006). According to these authors life emerged around hydrothermal vents in connection with reactions involving H_2, HCOO-, CH_3S- and CO_2 catalyzed by sulphide, analogous to the synthesis of acetate ($H_3C.COO$-). In this model Glycine ($+H_3N. CH_2.COO$-) and other amino acids, as well as tiny quantities of RNA, were trapped within tiny iron sulfide cavities. The released energy from these reactions resulted in polymerizing glycine and other amino acids into short peptides upon the phosphorylated mineral surface. RNA acted as a polymerizing agent for amino acids as a process regulating metabolism and transferring genetic information. Such biosynthetic pathways probably evolved before 3.7 Ga when reduction prevented a buildup of free atmospheric oxygen (Russell and Hall 2006).

The oldest chemical manifestations of life discovered to date are in the form of dolomite and banded iron sediments (BIF) from 3.85 Ga-old Akilia banded ironstones and dolomite, southwest Greenland, where these high-grade metasediments have seawater-like Rare Earth element and Y (REE+Y_{SN}) signatures (Bolhar et al. 2004; Friend et al. 2007) (Fig. 2.9) consistent with those of sea water (Nutman et al. 2010), similar to metasediments in the 3.71–3.70 Ga-old Isua greenstone belt, southwest Greenland. Theories regarding the origin of BIF hinge on oxidation by microbial photoautotrophs or, alternatively, abiotic photo-oxidation of ferrous to ferric iron under the unoxidizing conditions of the early atmosphere/hydrosphere (Cloud 1973; Garrels et al. 1973; Konhausser et al. 2002). A biogenic origin of BIF is supported by evidence for fractionation of the Fe isotopes relative to Fe in igneous rocks (Dauphas et al. 2004). The significance of dolomite hinges on experimental studies indicating precipitation of low-temperature dolomite in sedimentary systems and interstices of pillow lava under unoxidizing conditions requires microbial mediation (Vasconcelos et al. 1995; Roberts et al. 2004). Alternative views regard the dolomite as the product of metasomatism (Rose et al. 1996). Previously low ^{13}C in graphite clouding in Akilia apatite were interpreted in terms of early biogenic activity (Schidlowski et al. 1979; Mojzsis et al. 1996; Rosing 1999). Questions raised in this regard concern the sedimentary origin of host rocks and inorganic de-carbonation processes (Perry and Ahmed 1977; van Zuilen et al. 2002). Further, the significance of low $^{12}C/^{13}C$ indices as discriminants between biogenic and non-biogenic processes has been questioned (McCollom and Seewald 2006).

A study of 3.47–3.30 Ga carbonaceous chert layers and veins in the Onverwacht Group of the Barberton Greenstone Belt suggests an origin of the bulk of the carbonaceous matter in the chert by biogenic processes, accompanied by modification through hydrothermal alteration (Walsh and Lowe 1985; Walsh 1992). These studies identified textural evidence in cherts for microbial activity represented by carbonaceous laminations in intercalations within the Hoogenoeg and Kromberg Formations, with affinity to modern mat-dwelling cyanobacteria or bacteria. These include a range of spheroidal and ellipsoidal structures analogous to modern coccoidal bacteria and bacterial structures, including spores. The Pilbara Craton, Western Australia, contains evidence of >3.0 Ga microfossils, trace fossils, stromatolites, biofilms, microbial and microscopic sulfide minerals with distinctive biogenic sulfur isotope signatures (Wacey 2012). Schopf and Packer (1987) identified eleven taxa, including eight new species of cellularly preserved filamentous Prokaryote microbes in a shallow chert sheet of the ~3.459 Ga Apex Basalt. This assemblage indicates morphologically diverse extant trichomic cyanobacterium-like microorganisms, suggesting presence of oxygen-producing photoautotrophs. However, a possibility remains these microbes represent contamination by ground water.

Studies of microbial remains from the ~3.49 Ga Dresser Formation, North Pole Dome, central Pilbara Craton (Glikson et al. 2008) (Fig. 11.11) and of filamentous microbial remains and carbonaceous matter (CM) from Archaean black cherts (Duck et al. 2007, 2008) (Fig. 11.12) have used the following methods: (1) organic petrology; (2) Transmission Electron Microscopy (TEM); (3) Electron Dispersive Spectral Analysis (EDS); (4) high resolution TEM (HRTEM); (5) elemental and carbon isotope geochemistry studies (6) reflectance measurements determining thermal stress. The analyses resolve images of microbial relics and cell walls analogous to the modern hyperthermophilic Methano-caldococcus jannaschii residing in hydrothermal sea floor environments. Analogies include the wall structure and thermal degradation mode about 100 °C considered as the upper limit of life, whereas complete disintegration takes place at 132 °C. The $\delta^{13}C$ values of CM from the ~3.49 Ga Dresser Formation (−36.5 to −32.1‰) show negative correlation with total organic carbon (TOC = 0.13–0.75 %) and positive correlation with Carbon/Nitrogen ratios (C/N = 134–569). These values are interpreted in terms of oxidation and recycling of the CM and loss of light ^{12}C and N during thermal maturation. The TEM and carbon isotopic compositions are consistent with activity of chemosynthetic microbes in a seafloor hydrothermal system accompanied with rapid silicification at relatively low temperature.

Transmission Electron Microscopy (TEM) studies of filamentous and tubular structured isotopically light ($\delta^{13}C$ −26.8 to −34.0‰ V-PDB) carbonaceous material associated with ~3.24 Ga epiclastic and silicified sediments overlying sulphide (Fig. 11.12), indicate close analogies with sea floor hydrothermal environments (Duck et al. 2007). The total organic carbon (<1.0–2.3 %) and thermal maturity obtained by reflectance (%Ro) indicates maximum temperatures around 90–100 °C. The association of sulphide with the organic matter suggests a sediment-hosted microbial community and seafloor hydrothermal activity.

Isotopic sulphur $\delta^{33}S$ values from the ~3.49 Ga Dresser Formation and ~3.24 Ga Panorama Zn–Cu deposit indicate presence of seawater sulfate and elemental sulfur of UV-photolysis origin (Golding et al. 2011), representing mixing between mass independently fractionated sulfur reservoirs with positive and negative $\delta^{33}S$ in the Dresser Formation. Pyrite associated with barite is depleted in $\delta^{34}S$ relative to the host barite, interpreted as evidence for microbial sulfate reduction. Alternatively the pyrite may have formed by thermochemical reaction inferred for the chert-barite assemblage units in excess of 100 °C. For the ~3.24 Ga deposit the data absence of significant negative $\delta^{33}S$ anomalies in sulphide suggests volcanic sulfur rather than seawater sulfate.

11.2 Archaean Life

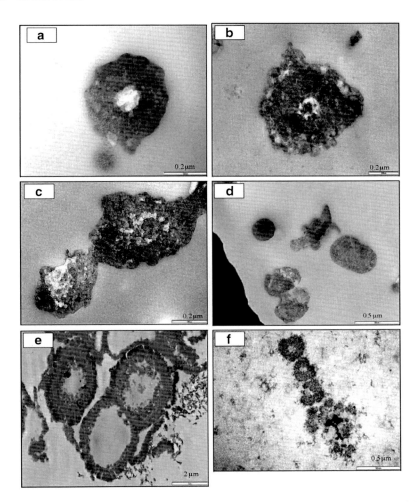

Fig. 11.11 Microbial remains from the ~3.49 Ga Dresser Formation. TEM micrographs of thermally degraded Carbonaceous Matter (*CM*) concentrate after demineralization of rock. (*A and B*) Single granular-textured bodies showing porosity and central cavity following dissolution of mineral matter (samples DF-17 (**a**); DF-10 (**b**)). (**c**) Micro-porous bodies connected by extensions of same CM (sample DF-10). (**d**) Cell-like bodies 'caught' at possible early stages of dividing or splitting (sample DF-17). (**e**) Part of an aggregate of cell-like bodies showing central cavities as characteristic of CM concentrate after demineralization (sample H-1). (**f**) M. jannaschii harvested after being autoclaved at 132C, showing nano-porosity following breakdown and ejection of cellular material. Shrinkage, deformity of cells and amorphous material formation following thermal stress is evident as is contraction of cell cavity into a tiny central hollow (From Glikson et al. 2008; Elsevier, by permission)

The observation of stromatolite-like forms in chert-carbonate-barite sequence of the ~3.49 Ga Dresser Formation, North Pole dome (Fig. 8.5), central Pilbara Craton (Dunlop et al. 1978; Dunlop and Buick 1981) (Fig. 11.6) gave rise to a major controversy regarding the biogenic origin of these structures. Noffke et al. (2013)

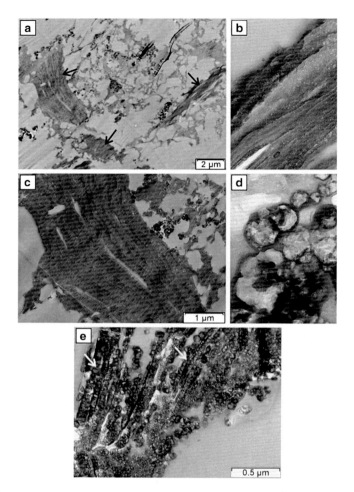

Fig. 11.12 Microbial remains from the 3.24 Ga Sulphur Springs black smoker deposit (**a**) TEM image illustrating the filamentous structure of the striated OM (*arrowed*). This material is comprised of compressed filaments and bundles of tubular structures, ranging from 1–5 μm in width and up to 100 μm in length; (**b**) Higher magnification TEM image of compressed filaments illustrated in (**a**). (**c**) TEM surface view of bundles of tubular microbial remains; (**d**) TEM cross section of a tubular bundle; (**e**) TEM image of silicified microbial tubes (From Duck et al. 2007; Elsevier, by permission)

documented microbially induced sedimentary structures from the Dresser Formation and interpreted the relations between microbial mats and the physical sedimentary environment. Detailed mapping on the scale of meter to millimeter indicates five sub-environments typical of coastal sabkha which contain distinct macroscopic and microscopic associations of microbially induced sedimentary structures. Outcrop-scale microbial mats include polygonal oscillation cracks and gas domes, erosional remnants and pockets, and mat chips. Microscopic microbial lamina comprise tufts,

11.2 Archaean Life

sinoidal structures, and lamina fabrics and consist of primary carbonaceous matter, pyrite, and hematite, plus trapped grains.

Greater confidence exists regarding the nature of the 3.43 Ga Strelley Pool Chert (Lowe 1980) (Figs. 11.7 and 11.8). Despite reservations (Brasier et al. 2002; Lindsay et al. 2003) the heliotropic reef-forming structure of the microbialites supports a biological ooorigin (Hoffman et al. 1999; Van Kranendonk et al. 2003). Alwood et al. (2006a, b, 2007) documented evidence for palaeoenvironmental extensive stromatolite-like reefs, including seven stromatolite morphotypes in different parts of a peritidal carbonate platform. The diversity, complexity and environmental associations of the stromatolites display marked analogies to similar stromatolite reef settings in younger geological systems.

Late Archaean ~2.73 Ga stromatolites, containing inter-bioherm interstitial debris are widespread in the Fortescue Basin, Pilbara Craton, reaching dimensions of tens of meters (Fig. 11.9), and yet younger ~2.63 Ga stromatolites have flourished following the JIL impact event in the central and East Fortescue Basin (Fig. 11.10).

Chapter 12
Uniformitarian Theories and Catastrophic Events Through Time

Abstract Uniformitarian models for the early Earth take little or no account of repeated impacts of asteroid clusters and their effects on crust and mantle. However a large body of evidence exists for multiple impacts by bodies on the scale of tens of kilometer during ~3.47–2.48 Ga (Lowe et al. Astrobiology 3:7–48, 2003; Lowe and Byerly, Did the LHB end not with a bang but with a whimper? 41st Lunar Planet Science conference 2563pdf, 2010; Glikson and Vickers. Aust J Earth Sci 57:79–95, 2010; Glikson The asteroid impact connection of planetary evolution. Springer-Briefs, Dordrecht, 150 pp, 2013), likely accounting at least in part for mafic-ultramafic volcanism produced by mantle rebound and melting events, consistent with original suggestion by Green (Earth Planet Sci Lett 15:263–270, 1972; Green DH Petrogenesis of Archaean ultramafic magmas and implications for Archaean tectonics. In: Kroner A (ed) Precambrian plate tectonics. Elsevier, Amsterdam, pp 469–489, 1981). Further, the juxtaposition of at least four impact ejecta units with the fundamental unconformity between granite-greenstone terrains and semi-continental deposits in both the Barberton Greenstone Belt and the Pilbara Craton about ~3.26–3.227 Ga constitutes a primary example for the tectonic and magmatic effects of asteroid impact clusters in the Archaean, supporting Lowe and Byerly's (Did the LHB end not with a bang but with a whimper? 41st Lunar Planet Science conference 2563pdf, 2010) suggested extension of the late heavy bombardment (LHB).

Keywords Evolution • Uniformitarianism • Catastrophism • Crustal evolution • Late heavy bombardment

James Hutton (1726–1797), observing the great thickness of sedimentary successions, deduced the long-term development of erosion and deposition over a vast expanse of geological time, a concept philosophically related to Darwin's evolutionary theory which has progressively replaced the theory of catastrophism, which advocated violent short-lived events outside our present experience drastically affected the earth. The new trend culminated with Lyell's (1797–1875) uniformitarian doctrine, suggesting the natural laws and geological processes observed in the present time have operated in the past, as summed up by the dictum: *"the present is the key to the past"*. According to Hutton's (1788) principle *"no powers are to be employed that are not natural to the globe, no action to be admitted except those of which we know"*. Hutton insisted that the way and means of nature could be discovered only by observation, an empirical approach emphasized by Bullard (1964) who stated *"it is usually best to decide on the existence of a phenomenon in the light of the known facts, or of a reasonable explanation of them, and then to look for a mechanism or a physical theory"*. Holmes (1965) recognized this problem and regarded uniformitarianism as an unhappy word liable to be taken too literally, stating: *"Lyell's term inevitably suggests a uniformity of rate, whereas what is meant is a uniformity of natural laws"*.

Mitroff (1974), investigating the methodological questions associated with NASA's lunar rocks study, made the observation that, in multi-disciplinary studies, a proliferation of models is an inherently necessary development since usually no single mind does objectively master, assess and verify all the evidence for several contradictory hypotheses. For this reason workers supporting what are eventually shown to be mistaken concepts render as important a service to science as those who prove correct, as in this way the truth should come out in the wash. Where the conditions for debate do not exist and basic scientific ethics are not observed, dominant dogmas take over scientific progress. Any investigation of complex natural phenomena requires that both the limitations of individual methods and their implications to one another are recognized. It is common practice for workers to extrapolate observations from a single terrain or a single method into other areas or subjects.

The empirical methodology inherent in the development of plate tectonic theory suggests its extension into earlier chapters of earth history constitutes a preconceived uniformitarian assumption. By contrast, multidisciplinary analysis and synthesis of direct observations, adhering to a parsimony principle, is capable of constraining concepts on the spatial-temporal distribution of crustal segments. Where more than one explanation exists for a set of observations, the one with the least number of unknowns ought to be tentatively preferred. However, the rapidly growing Archaean databases over the last 50 years or so, rather than constraining geotectonic theories, has led to a diversification of models, including models unconstrained by direct observations and geochemical and isotopic data – constituting a fundamental methodological impasse.

Following the demonstration of plate-tectonics it has become commonplace to interpret a range of features in Precambrian terrains, including structural lineaments, faults, shears, magnetic and gravity lineaments, elongated intrusions, volcanic belts, metamorphic isograds, geochemical anomalies, in terms of plate tectonic elements compared to arc-trench, back-arc, Cordillera systems, or intra-plate settings or mantle plumes (cf. Pearce and Cann 1971, 1973; Pearce et al. 1977; Katz 1972; Dewey and Spall 1976; Tarney et al. 1976; Windley and Smith 1976; Walker 1976; Tarney and

Windley 1977; Burke et al. 1976; Kroner 1981, 1991; de Wit 1998; Krapez and Eisenloh, 1998; Kerrich and Polat 2006; Pirajno 2000, 2007a, b; Krapez and Barley 2008; Percival et al. 2006; Burke, 2011; Jenner et al. 2006, 2013; Polat 2013; Beresford et al. 2013). Inherent in uniformitarian models of Archaean terrains is the assumption that similar petrogenetic processes and trace element patterns necessarily imply analogous geotectonic environments. Based on extensive observations in Mesozoic-Cenozoic circum-Pacific belts, Hamilton (2003) concluded Archaean crustal regimes are fundamentally distinct from subduction-related orogenic belts and accretion wedges. Thus, no ophiolite and melange-type thrust-associated assemblages such as are diagnostic of circum-Pacific wedges have to date been identified in Archaean greenstone belts. The evidence for large asteroid clusters and their consequences (Chaps. 7, 8 and 10) is inconsistent with tectonic models based on purely internal/endogenic processes. Whereas the bulk of geochemical and isotopic data from early to mid-Archaean terrain displays similarities with modern systems underpinned by two-stage mantle melting processes (Green and Ringwood 1967, 1977; Green 1981), no reason has been given why similar petrogenetic processes cannot occur in distinct tectonic environments.

Moorbath (1977) proposed a view of crustal evolution in terms of "episodic uniformitarianism", i.e. intermittent operation of two-stage mantle melting processes leading to a series of major accretion-differentiation events in environments akin to modern circum-Pacific domains, where juvenile mantle-derived additions to continental crust dominate over intra-crustal reworking. However, whereas similarities exist between Archaean granite-greenstone terrains, Sierra Nevada Palaeozoic batholiths (Hietanen 1975) and Chilean Mesozoic granodiorite-greenstone back-arc terrains (Tarney et al. 1976), there is no evidence for the existence in the Archaean of large continents such as those fringing the circum-Pacific arc-trench belts. Inherent in plate-tectonic processes is subduction and partial melting of large volumes of oceanic crust and production of mafic to felsic magma in overlying mantle wedges, giving rise to belts of andesite, Na-rich dacite, Na-rich rhyolite and minor shoshonite and their dioritic, tonalitic and trondhjemitic plutonic counterparts. Early Archaean greenstone belts are dominated by bimodal mafic-felsic volcanic assemblages and plutonic counterparts are characterized by low initial $^{87}Sr/^{86}Sr$ values, heavy REE-depletion, light REE-enriched patterns, high Na/K and low LIL (Large ion lithophile) elements (Glikson 1979a). Andesite is a relatively rare component in pre-3.2 Ga volcanic sequences but common in younger greenstone belts (Superior Province, Midlands belt in Zimbabwe, Marda belt in Western Australia). These compositions are commonly distinct from those of several early and mid-Proterozoic terrains dominated by eutectic K-rich granites, adamellite, granodiorite and their extrusive equivalents, characterized by intermediate to high $^{87}Sr/^{86}Sr$ values, high LIL-element levels and negative Eu anomalies (Glikson 1976a, b).

Temporally unique features of the Archaean Earth include (Glikson 1980; Hamilton 2003):

1. The evidence for major asteroid impact clusters at several stages of Archaean and Early Proterozoic history, including ~3,482, 3,472, 3,445, 3,416, 3,334, 3,256, 3,243, 3,225 (impact ejecta units), 2,975 (Maniitsoq, Greenland), 2,630, 2,570, 2,560, 2,481, 2,020 (Vredefort), 1,850 (Sudbury) Shoemaker and Shoemaker 1996; French 1998; Glikson and Vickers 2010; Glikson 2013).
2. In particular impact, tectonic and magmatic records for the interval ~3,256–3,225 Ma in the Barberton Greenstone Belt and in the Pilbara Craton indicate major asteroid impact cluster resulting in faulting, uplifts, formation of unconformities and granite intrusions affecting abrupt shifts from mafic-ultramafic-TTG (tonalite-trondhjemite-granodiorite)-dominated crust to continental crust (Lowe et al. 2003; Glikson and Vickers 2006, 2010; Glikson 2013).
3. Evidence for high-temperature melting producing high-Mg and peridotitic komatiites (Green 1981; Arndt et al. 1997).
4. Trace element features suggesting presence of garnet in residues of partial melting, based on rare Earth element fractionation (Hanson 1975; Arth and Hanson 1975; Arth 1976); Modern arc-trench volcanics generally display lesser degrees of REE fractionation. High abundances of transition metals (Ni, Cr, Co) in Archaean andesites and some dacites, as compared to modern equivalents, suggest a high ultramafic component in the source of these magmas.
5. Presence of superheated tonalite-trondhjemite-granodiorite (TTG) magmas (Glikson and Sheraton 1972; Glikson 1979a, b). The degree to which mafic-ultramafic enclaves and xenoliths are digested by Archaean TTGs (Chap. 4) appears to exceed those observed in Phanerozoic batholiths.
6. Dominance of bimodal tholeiitic basalt-dacite volcanic suites and relative paucity of andesite in pre-3.2 Ga early to mid-Archaean suites (cf. Glikson and Hickman 1981);
7. An absence of Phanerozoic-type ophiolites-turbidite-flysch association (Glikson 1980; Hamilton 2003).
8. A paucity of alkaline igneous rocks, except as minor occurrences in late Archaean (cf. Smithies and Champion 1999).
9. Relatively little evidence of intra-continental anatectic processes during 3.6–3.2 Ga; Thus, in granite-greenstone systems K-rich granites mostly appear at a late stage of evolution of Archaean batholiths (cf. Robb and Anhaeusser 1983).
10. Dominance of vertical crustal movements in pre-3.2 Ga early to mid-Archaean terrains; commonly transcurrent and thrust deformation of these terrains occurred at upper or late Archaean stage.
11. Frequency distribution diagrams of isotopic ages plotted against time point to a strongly episodic nature of Archaean tectonic-thermal history (Fig. 12.1), interpreted in part as due to the tectonic and magmatic effects of large asteroid impacts (Glikson and Vickers 2010; Glikson 2013).

Analogies between early Archaean mafic-ultramafic belts and oceanic crust (Glikson 1971, 1972a, b) or Mare-like crust produced by magmatism triggered by asteroid impacts (Glikson 2005) point to an apparent absence of directly observed basal unconformities below basal mafic-ultramafic sequences and a granitoid basement.

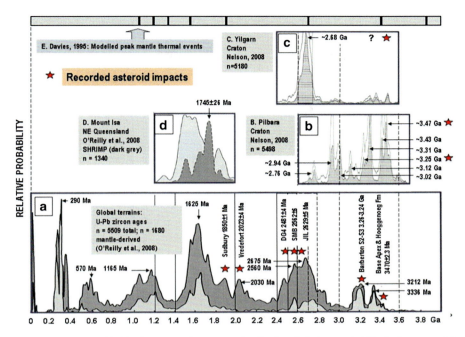

Fig. 12.1 Isotopic U–Pb zircon age frequency distribution (relative probability) diagrams. (**a**) Global (After O'Reilly et al. 2008). (**b**) Pilbara Craton, Western Australia (After Nelson 2008). (**c**) Yilgarn Craton, Western Australia (After Nelson 2008). (**d**) Northwest Queensland (After Nelson 2008). (**e**) Model mantle thermal events (Davies 1995). *Stars* represent recorded large asteroid impacts. Impact ages are located *above stars*; histogram peaks are indicated by *horizontal arrows* (Glikson and Vickers 2010; Australian Journal of Earth Sciences, by permission)

These sequences include the Onverwacht Group (~3.548–3.258 Ga; Kaapvaal Craton), Sebakwian Group (>3.5 Ga; Zimbabwe Craton) and the Warrawoona and Coonterunah Groups of the Pilbara Craton. In this view occurrences of entrained zircons and early ^{143}Nd/^{144}Nd model ages, indicating precursor granitic materials (Jahn et al. 1981; Smithies et al. 2007a; Van Kranendonk et al. 2007a, b; Hickman 2012), may reflect detrital components derived from adjacent older gneiss terrains. In the Pilbara Craton such older gneisses include the 3.66–3.58 Ga Warrawagine gneisses (Nelson 1999) and in the Kaapvaal Craton the Ancient Gneiss Complex (~3.7–3.5 Ga; Kröner et al. 1996; Kröner and Tegtmeyer 1994).

The intrusion into greenstone belts of tonalite-trondhjemite-granodiorite (TTG) magmas whose geochemical and isotopic features imply derivation by partial melting of basic materials, militating against a sub-greenstone continental basement, as anatexis of such basement during extensive TTG magmatism would produce differentiated K-rich magmas.

In the Pilbara Craton constraints on evolutionary theories include the >15–20 km cumulative thickness 3.53–3.23 Ga volcanic sequence of the greenstone belts, including at least 8 volcanic cycles each approximately 10–15 million years-long, as based on zircon ages of felsic volcanic units (Hickman 2012). These cycles, representing mantle and crust melting events, were compared to those of Large Igneous Provinces (LIPs) formed by mantle plumes (Arndt et al. 2001; Condie 2001; Ernst et al. 2004). However, Phanerozoic LIPs are not known to be intruded by TTG magmas. Plate tectonic analogies of Archaean evolution in terms of Phanerozoic plate tectonics processes (Bickle et al. 1983; Barley 1993; Zegers et al. 2001; Kitajima et al. 2001; Terabayashi et al. 2003; Smithies et al. 2003) focus on similarities of tectonic and magmatic elements but overlook the differences. Thus, comparisons between the Archaean mafic-ultramafic sequences and modern oceanic crust overlook the occurrence of felsic volcanic intercalations within ultramafic-mafic sequences (cf. 3,548±3–3,544±3 Ma felsic tuff within the Theespruit Formation, Barberton Greenstone Belt; felsic volcanics in the ~3,515–3,498 Ma Coucal Formation, Coonterunah Subgroup, Pilbara Craton, Fig. 8.3) and the presence of unconformities, such as the pre-3.43 Ga unconformity truncating the Coonterunah Subgroup (Buick et al. 1995).

An example of a possible intra-cratonic rift zone is furnished by the ~3.26–3.22 Ga–old Strelley greenstone belt, central Pilbara Craton, which is down-faulted between older 3.49–3.42 Ga terrains (Fig. 8.5). Where locally the Strelley Group unconformably overlaps the older terrain, these relationships do not necessarily imply the older terrain originally formed a continuous basement below the Strelley supracrustals, which could have been deposited in a tectonic rift zone. No sub-greenstones continental basement is required by seismic mid-crustal data (Swager et al. 1997) from Archaean terrains in the Yilgarn Craton. Here, major sub-horizontal shears and thrusts, which detach high level supracrustals from deep-level granitic sheets, preclude identification of the original field relations between these sheets. Thus the older gneiss terrains do not necessarily suggest sub-greenstones continental crust more than suggested, for example, by zircon-bearing Red Sea rift sediments. As another example, the continental rise of Lord Howe Island does not imply existence of felsic crust beneath the Tasman Sea.

Martin (1999) and Smithies (2000) suggest that the Archaean tonalite-trondhjemite-granodiorite suite (TTG – Glikson and Sheraton 1972; Glikson 1979a, b) can be distinguished from modern analogues (adakites) in terms of the low Mg# (Fig. 6.10), Ni and other trace metals and high SiO_2 values of the Archaean rocks, suggesting little interaction between the Archaean magmas and the mantle. Smithies et al. (2003) suggests that, in contrast to modern low-angle subduction zones where magmas form by melting of both oceanic crust and mantle wedge, little or no subduction occurred prior to about 3.1 Ga and that instead thick mafic crust was directly melted, a process they refer to as "flat subduction". Exceptions occur, for example tonalites of the eastern Kaapvaal Craton with high Ni levels, possibly reflecting partial melting of komatiites which are abundant in this terrain (Glikson 1976a, b, c).

The identification of asteroid impact ejecta/fallout units in Archaean sedimentary sequences offers an explanation for a number of major Archaean episodes, as follows (Glikson and Vickers 2006, 2010; Glikson 2013):

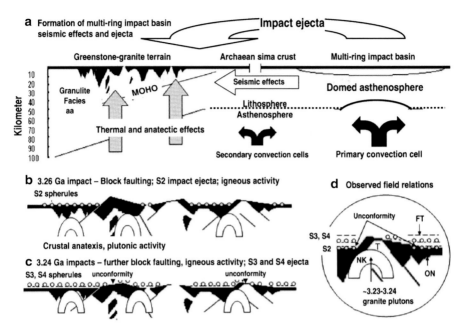

Fig. 12.2 A schematic model (not to scale) portraying the principal stages in asteroid impact-triggered cratering in mafic-ultramafic (*SIMA*) crustal regions of the Archaean Earth, ensuing rearrangement of mantle convection patterns, seismic activity, faulting, uplift, development of unconformities and igneous activity in both SIMA regions and affected greenstone-granite terrain. (**a**) 3.26 Ga (Barberton Greenstone Belt spherule unit S2 impact): formation of a multi-ring impact basin by a 20 km asteroid impact, impact ejecta, seismically triggered faulting, mantle convection underlying impact basin, secondary mantle cells, thermal and anatectic effects across the asthenosphere/lithosphere boundary and the MOHO below greenstone-granite nuclei. (**b**) 3.26 Ga: Block faulting in greenstone-granite nuclei, rise of anatectic granites, S2 spherules preserved in below-wave base environments. (**c**) 3.24 Ga – S3 and S4 impacts, ejecta fallout over below-wave base S2 ejecta and overlying sediments and over unconformities. Further faulting, block movements and rise of plutonic magmas. (**d**) Schematic representation of observed field relations between the ~3.55–3.26 Ga mafic–ultramafic volcanic Onverwacht Group (*ON*), intrusive early tonalites and trondhjemites (*T*), 3.26–3.24 Ga granites (*NK*), S2 ejecta, unconformity, S3, S4 ejecta, and the Fig Tree Group sediments (*FT*) (From Glikson, 2008; Elsevier, by permission)

~**3.47 Ga**. The association of a microkrystite spherules unit with at the top of the 3,470 ± 2 Ma felsic volcanics of the Antarctic Chert Member (uppermost Mount Ada Basalt) (Figs. 8.14 and 8.15) correlated with felsic volcanics of the Duffer Formation (3,471–3,473 Ma), hints at an association of crustal anatectic event with multiple asteroid impacts (Glikson et al. 2004). Recently a possible equivalent of this impact deposit was found below the Marble Bar Chert some 50 km to the east.

~**3.26–3.227 Ga**. The asteroid impact cluster identified in the Barberton Greenstone Belt (Lowe et al. 1989, 2003), with correlated units in the central Pilbara Craton, defines an abrupt transition from mafic-ultramafic crust to continental crust (Glikson and Vickers 2006, 2010; Glikson 2013) (Fig. 12.2) and a boundary between an early to mid-Archaean era and upper to late Archaean era. The significance of this period is underpinned by peak zircon ages of ~3265 Ma (~3.2–3.3 Ga) measured in the southwest Yilgarn Craton (Pidgeon et al. 2010) (Fig. 9.3). Whereas the former era is characterized by diapiric granitoids and vertical tectonics, the latter era is characterized by lateral accretion of elongate greenstone belts and thus by horizontal tectonics.

~2.63–2.48 Ga. Repeated asteroid clusters occur during the late Archaean, including major events at ~2.63, ~2.57, ~2.56 (double impact) and ~2.48 Ga. The relation between these impacts and late igneous activity in granite-greenstone terrains in the Yilgarn Craton (Figs. 9.1 to 9.4) and the late Archaean in the western Dharwar Craton, India (Mohan et al. 2014) are unclear.

To sum up, uniformitarian models take little or no account of repeated impacts of asteroid clusters and their effects on the Archaean Earth, despite of the large body of evidence for multiple impacts by bodies on the scale of tens of kilometer during ~3.47–2.48 Ga (Lowe et al. 2003; Lowe and Byerly 2010; Glikson and Vickers 2010; Glikson 2013), likely accounting in part for mafic-ultramafic volcanism produced by mantle rebound and melting events, consistent with original suggestion by Green (1972, 1981), and major breaks in crustal evolution such as about ~3.26–3.227 Ga.

Appendices

Appendix 1: U-Pb Zircon Ages (Except Where Specified Otherwise) from the Ancient Gneiss Complex, Swaziland (AGC) and Barberton Greenstone Belt, Eastern Transvaal (BGB) (Compiled After Poujol 2007)

East Kaapvaal Craton			
AGC NW Swaziland	U-Pb zircon in banded tonalite gneiss	3,644 ± 4 Ma 3,504 ± 6 Ma 3,433 ± 8 Ma	Compston and Kröner (1988)
AGC	U-Pb zircons in foliated to massive granodiorite	3,702 ± 1 Ma	Kröner et al. (1996) and Kröner and Tegtmeyer (1994)
AGC	U-Pb Ngwane gneiss	3,683 ± 10 Ma 3,521 ± 23 Ma 3,490 ± 3 Ma	Kröner et al. (1996) and Kröner and Tegtmeyer (1994)
BGB	Granite pebble in Moodies Fm	3,570 ± 6 Ma 3,518 ± 11 Ma 3,474 + 35/−31 Ma	Kröner et al. (1989) Tegtmeyer and Kröner (1987)
BGB	Zircon xenocryst in felsic volcanic Hoogenoeg Fm	3,559 ± 27 Ma	Kröner et al. (1989)
AGC	Foliated tonalitic gneiss	3,563 ± 3 Ma	Kröner et al. (1989)
BGB	Felsic volcanics, Theespruit Fm	3,548 ± 3–3,544 ± 3 Ma	Kröner et al. (1996)
BGB	Porphyritic granodiorite	3,540 ± 3 Ma	Kröner et al. (1996)
BGB	A tonalitic gneiss wedge in Theespruit Fm	3,538 + 4/−2 Ma 3,538 ± 6 Ma	Armstrong et al. 1990 and Kamo and Davis (1994)
BGB	Theespruit Fm felsic volcanics	3,531 ± 10 Ma	Armstrong et al. (1990)
	Theespruit Fm felsic volcanics	3,511 ± 3 Ma	Kroner et al. (1992)

(continued)

BGB	Steynsdorp pluton, banded trondhjemitic gneiss	3,510±4 Ma to 3,505±5 Ma	Kamo and Davis (1994)
BGB	Steynsdorp pluton, granodioritic phase	3,511±4 Ma 3,502±2 Ma	Kröner et al. (1996) Kamo and Davis (1994)
AGC	Trondhjemitic gneiss	3,504±24 Ma	Kröner et al. (1989)
BGB	Steynsdorp pluton, trondhjemitic gneiss	~3,490	Kröner et al. (1991a, b) and (1992)
BGB	Zircon xenocrysts in Steynsdorp pluton	3,553±4 Ma, 3,538±9 Ma 3,531±3 Ma	Kröner et al. (1991a, b) and (1996)
BGB	Stolzburg pluton, trondhjemite gneiss intrusive into Theespruit Fm	3,460+5/−4 Ma 3,445±3 Ma 3,431±11 Ma	Kamo and Davis (1994) Kröner et al. (1991a, b) Dziggel et al. (2006)
BGB	Quartz porphyry dyke, Komati Fm	3,470+39/−9 Ma 3,458±1.6 Ma (titanite age)	
BGB	Intercalation of tuff in Komati Fm	3,482±2 Ma	Dann (2000)
BGB	Meta-gabbro, Komati Fm	3,482±5 Ma	Armstrong et al. (1990)
BGB	Impact spherules	3,472 Ma Pilbara spherules: 3,470±2 Ma	Lowe and Byerly (2010)
AGC	Tonalitic gneiss	3,458±3 Ma	Kroner et al. (1989)
AGC	Tsawela Gneiss, a foliated tonalite	3,455–3,436 Ma	Kröner and Tegtmeyer (1994)
BGB	Trondhjemitic Theespruit pluton, intrudes the Sandspruit and Theespruit Fms	3,443–3,437 Ma	Armstrong et al. (1990), Kamo and Davis (1994), Kamo et al. (1990) and Kröner et al. (1991a, b, 1992)
BGB	Felsic tuff, Hoogenoeg Fm	3,445±6 Ma 3,438±12 Ma	deWit et al. (1987)
BGB	Impact spherules	3,445 Ma	Lowe and Byerly (2010)
BGB	Doornhoek pluton, trondhjemite gneiss, intrudes the Theespruit Fm.	3,448+4/−3 Ma	Kamo and Davis (1994)
	Reset titanite age	3,215 Ma (titanite)	
BGB	Theeboom pluton trondhjemite	3,460+5/−4 Ma	Kamo and Davis (1994)
	Titanite age (reset)	3,237–3,201 Ma	
BGB	Impact spherules	3,416 Ma	Lowe and Byerly (2010)
BGB	Lowest Kromberg Fm	3,416±7 Ma	Kröner et al. (1991a, b)
BGB	Komati Fm, pegmatitic gabbro	3,350 Ma	Armstrong et al. (1990) and Kamo and Davis (1994)
BGB	Stentor pluton	3,347+67/−60 Ma	Tegtmeyer and Kröner (1987)
BGB	Impact spherules	3,334 Ma	Lowe and Byerly (2010)
South-central Swaziland	Granodiorite, Usuthu suite	3,306±4 Ma	Maphalala and Kröner (1993)

(continued)

BGB	Top Kromberg Fm (Footbridge Chert)	3,334 ± 3 Ma	Byerly et al. (1996)
BGB	Mendon Fm	~3,245–3,330 Ma	Lowe and Byerly (2007)
BGB	Mendon Fm volcanics (Felsic tuff)	3,298 ± 3 Ma	Byerly et al. (1996)
BGB	Zircon xenocrysts, volcanic rocks, Fig Tree Group	3,334–3,310 Ma	Byerly et al. (1996) and Kröner et al. (1991a, b)
BGB	Granite pebble, Moodies Group	3,306 + 65/−57 Ma	Tegtmeyer and Kröner (1987)
BGB	Felsic tuff, base Fig Tree Group	3,258 ± 3 Ma	Byerly et al. (1996)
BGB	Fig Tree Group felsic volcanics, clastics and BIF	3,259–3,225 Ma	Armstrong et al. (1990), Byerly et al. (1996), Kamo and Davis (1994), Kohler et al. (1993) and Kröner et al. (1991a, b, 1992)
BGB	Felsic tuff	3,256 ± 4 Ma	Kröner et al. (1991a, b)
BGB	Impact spherules	3,256 Ma	Lowe and Byerly (2010)
BGB	Felsic tuff	3,253 ± 2 Ma	Byerly et al. (1996)
BGB	Impact spherules	3,243 Ma	Lowe and Byerly (2010)
BGB	Felsic tuff, base Fig Tree Group	3,243 ± 4 Ma	Kröner et al. (1991a, b)
BGB	Felsic tuff	3,227 ± 4 Ma	Kröner et al. (1991a, b)
BGB	Porphyry intrusive into the Fig Tree Group	3,227 ± 3 Ma	de Ronde (1991)
BGB	Kaap Valley pluton (tonalite)	3,229–3,223 Ma	Armstrong et al. (1990), Kamo and Davis (1994), Layer et al. (1992) and Tegtmeyer and Kröner (1987)
BGB	Kaap Valley pluton (tonalite) 40 Ar/39 Ar hornblende and biotite ages	3,214 and 3,142 Ma	Layer et al. (1992)
BGB	Nelshoogte pluton; U-Pb	3,236 ± 1 Ma	Kamo and Davis (1994)
BGB	Nelshoogte pluton; Pb-Pb zircon	3,212 ± 2 Ma	York et al. (1989)
BGB	Nelshoogte pluton; 40 Ar/39 Ar ages	3,080–2,860 Ma	York et al. (1989)
BGB	Stentor pluton (central part) – trondhjemite to granodiorite gneiss	3,250 ± 30 Ma	Tegtmeyer and Kröner (1987)
BGB	Stentor pluton (eastern part)	3,107 ± 5 Ma	Kamo and Davis (1994)
BGB	Dalmein pluton, granodiorite	3,216 + 2/−1 Ma	Kamo and Davis (1994)
BGB	Badplaas domain gneisses	3,290–3,240 Ma	Kisters et al. (2006)
NW Swaziland	Wyldsdale pluton	3,234 + 17/−4 Ma	Fletcher (2003)

(continued)

NC Swaziland	Usuthu suite, granodiorite	3,231 ± 4 Ma–3,224 ± 4 Ma	Maphalala and Kröner (1993)
AGC	K-feldspar-rich granitic gneiss	3,227 ± 21 Ma	Kröner et al. (1989)
AGC	Tonalitic dyke	3,229 Ma	de Ronde and Kamo (2000)
BGB	Titanite age	3,229 ± 25 Ma	Diener et al. (2005)
Tjakastad schist belt	Titanite from a sedimentary unit	3,229 ± 9 Ma	
	Youngest group of Tjakastad belt	3,227 ± 7 Ma	
	Late trondhjemite	3,229 ± 5 Ma	
BGB	Dacite, top Fig Tree Group	3,225 ± 3	Kröner et al. (1991a, b)
BGB	Impact spherules (at least 2 layers)	3,225 Ma	Lowe and Byerly (2010)
BGB	Pebble, Moodies Group	3,224 ± 6 Ma	Tegtmeyer and Kröner (1987)
BGB	Porphyry intrusion, Stolzburg Syncline	3,222 + 10/−4 Ma	Kamo and Davis (1994)
AGC	Tonalitic gneiss	3,214 ± 20 Ma	Kröner et al. (1989)
BGB	Granodiorite intrusive, Dwazile greenstone belt	3,213 ± 10 Ma	Kröner and Tegtmeyer (1994)
BGB	Moodies Group – older age limit	3,224 ± 6 Ma	Tegtmeyer and Kröner (1987)
BGB	Felsic dyke intruding Moodies Group	3,207 ± 2 Ma	Heubeck and Lowe (1994)
BGB	Moodies Group – younger age limit (Salisbury Kop pluton)	3,109 + 10/−8 Ma	Kamo and Davis (1994)
South-eastern Kaapvaal Craton			
	Layered light and dark TTG gneisses and amphibolite	3,640–3,460	Hunter et al. (1992)
	Luneburg gneisses, tonalitic to trondhjemitic gneiss	3,458–3,362 Ma	Hunter et al. (1992)
	Witkop formation of the Nondweni Group, zircons from a rhyolite flow	3,406 ± 6 Ma	Versfeld and Wilson (1992)
	Commondale greenstone belt, peridotite, Sm-Nd age	3,334 ± 18 Ma	Wilson and Carlson (1989)
	Mvunyana Granodiorite – Rb-Sr, Pb-Pb and limited U-Pb zircon	0.3290 Ma	Matthews et al. (1989)
	De Kraalen gneiss, layered tonalitic rocks	Intruded by 3,250 Ma Anhalt Granitoid Suite	Hunter et al. (1992)

(continued)

	Anhalt Granitoid Suite – trondhjemites, granodiorites and quartz monzonite.	Rb-Sr age 3,250 ± 39 Ma	Hunter et al. (1992) Farrow et al. (1990)
	Komatiite basalt flows, Re-Os isochron regression age	3,321 ± 62 Ma	Shirey et al. (1998)
	Natal Spa granite	3,210 ± 25 Ma	Reimold et al. (1993)
Northern Kaapvaal Craton			
	Zircon xenocrysts, Mac Kop conglomerate, Murchison greenstone belt	3,364 ± 18 Ma	Poujol et al. (1996)
	Goudplaats gneisses – migmatitic tonalitic gneiss	3333.3 ± 5 Ma	Brandl and Kröner (1993)
	Tonalitic gneisses	3282.6 ± 0.4, 3,274 + 56/ −45 Ma	Kröner et al. (2000)
	Felsic metavolcanic, Giyani greenstone belt	3203.3 ± 0.2 Ma	Brandl and Kröner (1993)
	Makhutswi Gneiss, tonalite to granodiorite	3,228 ± 12 Ma	Poujol et al. (1996)
Central Kaapvaal Craton			
	Trondhjemite gneiss, Johannesburg Dome	3,340 ± 3 Ma	Poujol and Anhaeusser (2001)
	Zircon xenocrysts, Vredefort dome	3,425 Ma 3,310 Ma	Hart et al. (1999) Kamo et al. (1996)
	Zircon xenocryt	3,480 Ma	Armstrong et al. (1991)
	Zircon xenocryt	3,259 ± 9	Armstrong et al. (1991)
	Zircon xenocryt	3,230 ± 8 Ma	Armstrong et al. (1991)
	Tonalitic gneiss, Johannesburg dome	3199.9 ± 2 Ma	Poujol and Anhaeusser (2001)
Western Kaapvaal Craton			
	Zircon xenocryt, felsic schist, Madibe greenstone belt	3,428 ± 11, 3,201 ± 4 Ma	Hirner (2001)
	Tonalitic and trondhjemitic gneisses, 207Pb/206Pb age	~3,250 Ma	Drennan et al. (1990)
	Tonalitic gneiss, 207Pb/206Pb	3,246–3,069 Ma	Schmitz et al. (2004)

Appendix 2: Key Supracrustal and Plutonic Units and Isotopic Ages, Pilbara Craton, Western Australia

Age (million years) (Ma)	Supracrustal unit	Lithology	Plutonic unit	Age (Ma)
West Pilbara Craton[a]				Karratha terrain
~2,940	Mallina formation	Siltstones, arenite, greywacke	~3.27–3.26 Ga T_{Nd} ~3.48–3.43 Ga	Karratha granodiorite
	Bookingarra formation	Siliceous high-Mg basalt	~2.93–3.27	Yule Batholith
	Constantine sandstone	Arenite, conglomerate	~2,945	Mungaroona granodiorite (high-Mg diorite)[c]
~2,950	Rushall Shale	Shale		
~2,960	Cistern formation	Fanglomerate, breccia and arenite		
~2,980	Cattle well formation	Felsic volcaniclastics		
	Coonieena Basalt	Basalt		
~3,000	Red hill volcanics	Volcanics		
	Warambie Basalt	Basalt		
	Cundaline formation	Conglomerate, arenite, shale		
~3,020	Cleaverville formation	Chert, BIF, carbonaceous shale, arenite		
	Farrel Quartzite	Arenite, conglomerate		
~3,118	Woodbrook formation	Tuff, agglomerate, basalt		
~3,120	Bradley Basalt	Pillowed basalt and komatiite		
~3,120	Tozer formation	Basalt, andesite, dacite, rhyolite and sediments		
~3,126	Nallana formation	Basalt, ultramafics, pyroclastic volcanics		
Sholl shear zone				
~3,190	Port Robinson Basalt	Banded iron formations		
	Dixon Island formation	Volcanics, felsic tuff		

(continued)

Regal thrust				
~3,180	Pilbara well volcanics	Mafic volcanics		
~3,180	Honeyeaster Basalt	Mafic volcanics		
	Paddy market formation	Submarine fan		
~3,190	Cardinal and Corboy formations	Clastic sediments		
~3,220	Budjan creek formation	Clastic sediments		
~3,250	Roebourne Group	Mafic volcanics		
~3,270	Ruth well formation	Mafic volcanics		
East Pilbara Craton				
~2,930	Mosquito creek formation	Siltstone, shale	~3.20–3.17	Kurrana terrain Gneiss granodiorite
~3,335	Kangaroo cave formation	Felsic volcanics and sedimentary intercalations	3,241 ± 3	Mount Edgar Batholith[b]
	Kunagunarinna formation	Mafic and ultramafic volcanics	3,243 ± 4	"
~3,270	Leilira formation	Arenite and conglomerate	3,294 ± 5	"
Unconformity			3,298 ± 2	"
	Charteris Basalt	Basalt	3,303 ± 6	"
~3,320	Wyman formation	Rhyolite	3,308 ± 3	"
~3,350	Euro Basalt	Mafic and ultramafic volcanics	3,314 ± 6	"
Disconformity			3,321 ± 6	"
~3,420	Strelley Pool formation	Carbonate, arenite, stromatolites	3,429 ± 13	"
Regional unconformity			3,437 ± 5	"
~3,430	Panorama formation	Felsic volcanics	3,448 ± 8	"
	Apex Basalt	Mafic to ultramafic volcanics		
~3,465	Duffer formation	Felsic volcanics (dacite to rhyolite)		
3,470.1+/−1.9 Ma				
~3,470	Mount Ada Basalt	Pillowed and carbonated basalts		
~3,480	Dresser and McPhee formations	Felsic volcanics, arenites, barite, chert		
~3,490	North Star Basalt	Mafic volcanics		
	Double bar formation	Mafic volcanics		
~3,515	Coucal formation	Felsic and mafic volcanics		
~3,525	Table top formation	Mafic volcanics		

[a]Supracrustal unit ages after Hickman (2012)
[b]Mount Edgar granite data after Van Kranendonk et al. (2007a): secular tectonic evolution of the Archaean continental crust interplay between horizontal and vertical processes in the formation of the Pilbara Craton, Australia Terra Nova 19, 1–38
[c]Geological Survey of Western Australia Annual Review 1999–2000

About the Author

Andrew Y. Glikson, an Earth and paleo-climate scientist, studied geology at the University of Jerusalem and graduated at the University of Western Australia in 1968. He conducted geological studies of the oldest geological formations in Australia, South Africa, India and Canada, studied large asteroid impacts, including effects on the atmosphere, oceans and mass extinction of species. Between 1998 and 2014, this work included detailed studies of impact ejecta units in the Pilbara Craton, Western Australia, geochemical studies of impact ejecta units from South Africa, and the identification and study of several Australian buried impact structures, including Woodleigh, Gnargoo, Mount Ashmore, Talundilly, Warburton and Winton. Currently he is a Visiting Scientist in Geoscience Australia developing a catalogue and website of Australian confirmed and possible asteroid impact structures. Since 2005 he extended the studies of past mass extinctions to the effects of climate on human evolution, the discovery of fire and global warming. He was active in communicating nuclear issues and climate change evidence to the public and parliamentarians through papers, lectures and conferences.

References

Aldiss DT (1991) The Motloutse Complex and the Zimbabwe Craton/Limpopo Belt transition in Botswana. Precambrian Res 50:89–109

Allaart JH (1976) The pre-3760 m.y. old supracrustal rocks of the Isua area, central west Greenland, and the associated occurrence of quartz banded ironstone. In: Windley BF (ed) Early history of the Earth. Wiley, London, pp 177–190

Allegre CJ, Luck JM (1980) Osmium isotopes as petrogenetic and geological tracers. Earth Planet Sci Lett 48:148–154

Allegre CJ, Dupre CJ, Lewin E (1986) Thorium/uranium ratio of the Earth. Chem Geol 56:19–227

Allwood AC, Walter MR, Kamber BS, Marshall CP, Burch I (2006a) Stromatolite reef from the Early Archaean era of Australia. Nature 441:714–717

Allwood AC, Walter MR, Marshall CP (2006b) Raman spectroscopy reveals thermal palaeo-environments of c. 3.5 billion-year-old organic matter. Vib Spectrosc 41:190–197

Allwood AC, Walter MR, Burch IW, Kamber BS (2007) 3.43 billion-year-old stromatolite reef from the Pilbara Craton of Western Australia: ecosystem-scale insights to early life on Earth. Precambrian Res 158:198–227

Anhaeusser CR (1973) The evolution of the early Precambrian crust of southern Africa. Philos Trans R Soc Lond A273:359–388

Anhaeusser CR (1974) Early Precambrian rocks in the vicinity of the Bosmanskop syenite pluton, Barberton Mountain Land, South Africa (abstract). In: Geology and geochemistry of the oldest Precambrian rocks symposium, University of Illinois, Champaign

Anhaeusser CR, Robb LJ (1978) Regional and detailed field and geochemical studies of Archaean trondhjemitic gneisses, migmatites and greenstone xenoliths in the southern part of the Barberton Mountain Land, South Africa. In: Smith IEM, Williams JG (eds) Proceedings of the 1978 Archaean geochemistry conference, University of Toronto, Ontario, pp 322–325

Anhaeusser CR, Mason R, Viljoen MJ, Viljoen RP (1969) A reappraisal of some aspects of Precambrian shield geology. Geol Soc Am Bull 80:2175–2200

Appel PWU, Fedo CM, Moorbath S, Myers JS (1998) Recognisable primary volcanic and sedimentary features in a low-strain domain of the highly deformed, oldest known (ca. 3.7–3.8 Gyr) greenstone belt, Isua, Greenland. Terra Nova 10:57–62

Archibald NJ, Bettenay LF, Binns RA, Groves DI, Gunthorpe RJ (1978) The evolution of Archaean greenstone terrains Eastern Goldfields Province, Western Australia. Precambrian Res 6:103–131

Armstrong RL (1968) A model of the evolution of strontium and lead isotopes in a dynamic Earth. Rev Geophys 6:175–199

Armstrong RA, Compston W, de Wit MJ, Williams IS (1990) The stratigraphy of 3.5–3.2 Ga Barberton greenstone belt revisited: a single zircon microprobe study. Earth Planet Sci Lett 101:90–106

Armstrong RA, Compston W, Retief EA, Williams IS, Welke HJ (1991) Zircon ion microprobe studies bearing on the age and evolution of the Witwatersrand triad. Precambrian Res 53:243–266

Arndt NT, Nelson DR, Compston W, Trendall AF, Thorne AM (1991) The age of the Fortescue group, Hamersley basin Western Australia from ion microprobe zircon U-Pb results. Aust J Earth Sci 38:261–281

Arndt N, Albarede F, Nisbet EG (1997) Mafic and ultramafic magmatism. In: de Wit MJ, Ashwal LD (eds) Greenstone belts. Oxford University Press, Oxford, pp 231–254

Arndt N, Bruzak G, Reischmann T (2001) The oldest continental and oceanic plateaus: geochemistry of basalts and komatiites of the Pilbara Craton Australia. In: Ernst RE, Buchan KL (eds) Mantle plumes: their identification through time. Geological Society of America (GSA) special publication 352. Geological Society of America, Boulder, pp 359–387

Arth JG (1976) Behavior of trace elements during magmatic processes: a summary of theoretical models and their application. J Res U S Geol Surv 4:41–47

Arth JG, Barker F (1976) Rare earth partitioning between hornblende and dacitic liquid and implication for the genesis of trondhjemitic–tonalitic magmas. Geology 4:534–536

Arth JG, Hanson GN (1975) Geochemistry and origin of the Early Precambrian crust of Northeastern Minnesota. Geochim Cosmochim Acta 39:325–362

Arth JG, Barker F, Peterman ZE, Fridman I (1978) Geochemistry of the gabbro-diorite-tonalite-trondhjemite suite of south-west Finland and its implications for the origin of tonalitic and trondhjemitic magmas. J Petrol 19:289–316

Baadsgaard H, Nutman AP, Bridgwater D, McGregor VR, Rosing M, Allaart JH (1984) The zircon geochronology of the Akilia association and the Isua supracrustal belt, West Greenland. Earth Planet Sci Lett 68:221–228

Baldwin RB (1985) Relative and absolute ages of individual craters and the rates of infalls on the Moon in the post-Imbrium period. Icarus 61:63–91

Ballhaus CG, Glikson AY (1995) The petrology of the layered mafic/ultramafic Giles complex, Western Musgrave Block, Western Australia. AGSO J Aust Geol Geophys 16:69–90

Baragar WRA, Goodwin AM (1969) Andesites and Archaean volcanism in the Canadian Shield. Oregon Dept Geol Miner Ind Bull 65:121–142

Baragar WRA, McGlynn JC (1976) Early Archaean basement in the Canadian Shield: a review of the evidence. Geological Survery of Canada Paper 14. Geological Survey of Canada, Ottawa

Barker F (1979) Trondhjemite: a definition, environment and hypotheses of origin. In: Barker F (ed) Trondhjemites, dacites and related rocks. Elsevier, Amsterdam, pp 1–12

Barley ME (1993) Volcanic, sedimentary, and tectono-stratigraphic environments of the ~3.46 Warrawoona Megasequence: a review. Precambrian Res 60:47–67

Barley ME, Eisenlohr BN, Groves DI, Perring CS, Vearncombe JR (1989) Late Archaean convergent margin tectonics and gold mineralization: a new look at the Norseman–Wiluna Belt, Western Australia. Geology 17:826–829

Barton JM (1981) The pattern of Archaean crustal evolution in southern Africa as deduced from the evolution of the Limpopo mobile belt and the Barberton granite-greenstone terrain. Geol Soc Aust Spec Publ 7:21–32

Beerling DJ, Berner RA (2005) Feedbacks and the coevolution of plants and atmospheric CO2. Proc Natl Acad Sci U S A 102:1302–1305

Bell CK (1971) Boundary geology, Upper Nelson River area, Manitoba and northwestern Ontario. Geol Assoc Can Spec Pap 9:11–39

Bell EA, Harrison TM, McCulloch MT, Young ED (2011) Early Archean crustal evolution of the Jack Hills Zircon source terrain inferred from Lu–Hf, ^{207}Pb/^{206}Pb, and δ^{18}O systematics of Jack Hills zircons. Geochim Cosmochim Acta 17:4816–4829

References

Beresford S, Tyler I, Smithies H (eds) (2013) Evolving early Earth. Precambrian Res 229:1–202

Berner RA (2004) The Phanerozoic carbon cycle: CO2 and O2. Oxford University Press, New York

Berner RA (2006) GEOCARBSULF: a combined model for Phanerozoic atmospheric O2 and CO2. Geochim Cosmochim Acta 70:5653–5664

Bettenay LF, Bickle MJ, Boulter CA, Groves DI, Morant P, Blake TS, James BA (1981) Evolution of the Shaw Batholith – an Archaean granitoid-gneiss dome in the east Pilbara, Western Australia. Geol Soc Aust Spec Publ 7:361–372

Beukes NJ, Gutzmer J (2008) Origin and paleoenvironmental significance of major iron formations at the Archean-Paleoproterozoic boundary. Soc Econ Geol Rev 15:5–47

Bickford ME, Wooden JL, Bauer RL (2006) SHRIMP study of zircons from Early Archaean rocks in the Minnesota River Valley: implications for the tectonic history of the Superior Province. Bull Geol Soc Am 118:94–108

Bickle MJ, Nisbet EG (1993) The geology of the Belingwe greenstone belt, Geological Society of Zimbabwe special publication. AA Balkema, Rotterdam, 239 pp

Bickle MJ, Martin A, Nisbet EJ (1975) Basaltic and peridotitic komatiites and stromatolites above a basal unconformity in the Belingwe greenstone belt, Rhodesia. Earth Planet Sci Lett 27:155–162

Bickle MJ, Bettenay LF, Barley ME, Chapman HJ, Groves DI, Campbell IH, de Laeter JR (1983) A 3500 Ma plutonic and volcanic calc-alkaline province in the Archaean East Pilbara Block. Contrib Mineral Petrol 84:25–35

Binns RA, Gunthorpe RJ, Groves DI (1976) Metamorphic patterns and development of greenstone belts in the Eastern Yilgarn Block, Western Australia. In: Windley BF (ed) Early history of the earth. Wiley, London, pp 331–350

Black LP, Gale N, Moorbath S, Pankhurst RT, McGregor VR (1971) Isotope dating of very early Precambrian amphibolite facies from the Godthab District, west Greenland. Earth Planet Sci Lett 12:245–259

Blais S, Auvrey B, Capdevilla R, Jahn BM, Hammeurt J, Bertrand JM (1978) The Archaean greenstone belts of Karelia and the komattitic and tholeiitic series. In: Windley, Naqvi (eds) Archaean geochemistry. Elsevier, Amsterdam, pp 87–108

Blake TS (1993) Late Archaean crustal extension, sedimentary basin formation, flood basalt volcanism, and continental rifting. The Nullagine and Mount Jope Supersequences, Western Australia. Precambrian Res 60:185–241

Bleeker W, Stern R (1997) The Acasta gneisses: an imperfect sample of Earth's oldest crust. In: Cook F, Erdmer P (eds) Slave-Northern Cordillera Lithospheric Evolution (SNORCLE) transect and Cordilleran tectonics workshop meeting. Lithosphere report 56, pp 32–35

Blenkinsop T, Martin A, Jelsma, HA, Vinyu ML (1997) The Zimbabwe Craton. In: de Wit M, Ashwal L (eds) Tectonic evolution of greenstone belts. Oxford University monographs on geology and geophysics 35. pp 567–580

Bohor BF, Glass BP (1995) Origin and diagenesis of K/T impact spherules – from Haiti to Wyoming and beyond. Meteoritics 30:182–198

Bolhar R, Kamber BS, Moorbath S, Fedo CM, Whitehouse MJ (2004) Characterisation of early Archaean chemical sediments by trace element signatures. Earth Planet Sci Lett 222:43–60

Bowring SA, Housh TB (1995) The Earth's early evolution. Science 269:1535–1540

Bowring SA, Williams IS (1999) Priscoan (4.00–4.03 Ga) orthogneisses from NW Canada. Contrib Mineral Petrol 134:3–16

Brandl G, Kröner A (1993) Preliminary results of single zircon studies from various Archaean rocks of the Northeastern Transvaal. In: Geological survey and mines, 16th colloquium of African geology, Mbabane, Swaziland, pp 54–56

Brasier MD, Green OR, Jephcoat AP, Kleppe AK, Van Kranendonk MJ, Lindsay JF, Steele A, Grassineau N (2002) Questioning the evidence for Earth's oldest fossils. Nature 416:76–81

Brauhart CW, Groves DI, Morant P (1998) Regional alteration systems associated with volcanogenic massive sulfide mineralization at Panorama, Pilbara, Western Australia. Econ Geol 93:292–302

Brazier MD, Green OR, Jephcoat AP, Kleppe AK, Van Kranendonk MJ, Lindsay JF, Steele A, Grassineau NV (2002) Questioning the evidence for Earth's oldest fossils. Nature 416:76–81

Bridgwater D, Collerson KD (1976) The major petrological and geochemical characteristics of the 3600 m.y. old Uivak Gneiss from Labrador. Contrib Mineral Petrol 54:43–60

Bridgwater D, Collerson KD (1977) On the origin of early Archaean gneisses. Contrib Mineral Petrol 62:179–190

Bridgwater D, McGregor VR, Myers JS (1974) A horizontal tectonic regime in the Archaean of Greenland and its implications for early crustal thickening. Precambrian Res 1:179–197

Buick R, Thornett JR, McNaughton NJ, Smith JB, Barley ME, Savage M (1995) Record of emergent continental crust ~3.5 billion years ago in the Pilbara Craton of Australia. Nature 375:574–577

Buick R, Brauhart CW, Morant P (2002) Geochronology and stratigraphic relationships of the Sulphur Springs Group and Strelley Granite: a temporally distinct igneous province in the Archaean Pilbara Craton, Australia. Precambrian Res 114:87–120

Bullard EC (1964) Continental drift. Q J Geol Soc Lond 120:1–19

Burke K (2011) Plate tectonics, the Wilson cycle, and Mantle plumes: geodynamics from the top. Ann Rev Earth Planet Sci 39:1–29

Burke K, Dewey JF, Kidd WS (1976) Dominance of horizontal movements, arcs and microcontinental collisions during the later premobile regime. In: Windley BF (ed) Early history of the Earth. Wiley, London, pp 113–130

Button A (1976) Transvaal and Hamersley Basins – review of basin development and mineral deposits. Miner Sci Eng 8:262–293

BVSP (Basaltic Volcanism on the Terrestrial Planets) (1981) Basaltic volcanism of the terrestrial planets. Pergamon, New York, 1286 pp

Byerly GR (1999) Komatiites of the Mendon Formation: late-stage ultramafic volcanism in the Barberton Greenstone Belt. In: Lowe DR, Byerly GR (eds) Geologic evolution of the Barberton Greenstone Belt, South Africa. Geological Society of America special papers 329. Geological Society of America, Boulder, pp 189–212

Byerly GR, Lowe DR (1994) Spinels from Archaean impact spherules. Geochim Cosmochim Acta 58:3469–3486

Byerly GR, Kröner A, Lowe DR, Todt W, Walsh MM (1996) Prolonged magmatism and time constraints for sediments deposition in the early Archaean Barberton greenstone belt: evidence from the Upper Onverwacht and Fig Tree Groups. Precambrian Res 78:125–138

Byerly GR, Lowe DR, Wooden JL, Xie X (2002) An Archaean impact layer from the Pilbara and Kaapvaal cratons. Science 297:1325–1327

Cairns-Smith AG (1978) Precambrian solution photochemistry, inverse segregation, and banded iron formations. Nature 276:807–808

Cameron WE, Nisbet EG, Dietrich VJ (1979) Boninites, komatiites and ophiolitic basalts. Nature 280:550–553

Campbell IH, Hill RI (1988) A two-stage model for the formation of the granite-greenstone terrains of the Kalgoorlie–Norseman area, Western Australia. Earth Planet Sci Lett 90:11–25

Canfield D, Poulton SW, Narbonne GM (2007) Late-neoproterozoic deep-ocean oxygenation and the rise of animal life. Science 315:92–95

Capdevila R, Goodwin AM, Ujike O, Gorton MP (1982) Trace-element geochemistry of Archean volcanic rocks and crystal growth in southwestern Abitibi Belt, Canada. Geology 10:418–422

Card KD (1990) A review of the Superior Province of the Canadian Shield, a product of Archaean accretion. Precambrian Res 48:99–156

Carmichael ISE, Turner FJ, Verhoogen J (1974) Igneous petrology. McGraw-Hill, New York

Cassidy KF, Champion DC, Krapež B, Barley ME, Brown SJA, Blewett RS, Groenewald PB, Tyler IM (2006) A revised geological framework for the Yilgarn Craton, Western Australia. Western Australia Geological Survey Record 2006/8. Geological Survey of Western Australia, Perth, 8 pp

References

Cavosie AJ, Wilde SA, Liu D, Weiblen PW, Valley JW (2004) Internal zoning and U-Th-Pb chemistry of Jack Hills detrital zircons: a mineral record of early Archean to Mesoproterozoic (4348–1576 Ma) magmatism. Precambrian Res 135:251–279

Cavosie AJ, Valley JW, Wilde SA (2005) Magmatic δ18O in 4400–3900 Ma detrital zircons: a record of the alteration and recycling of crust in the Early Archaean. Earth Planet Sci Lett 235:663–681

Cavosie AJ, Valley JW, Wilde SA (2007) The oldest terrestrial mineral record: a review of 4400 to 4000 ma detrital zircons from jack Hills, Western Australia. In: Van Kranendonk MJ, Smithies RH, Bennett VC (eds) Earth's oldest rocks, Developments in Precambrian geology 15. Elsevier, Amsterdam, pp 91–111

Cawood PA (2005) Terra Australis Orogen: Rodinia breakup and the development of the Pacific and Iapetus margins of Gondwana during the Neoproterozoic and Paleozoic. Earth Sci Rev 69:249–279

Chadwick B, Vasudev VN, Hegde GV (2000) The Dharwar craton, southern India, interpreted as the result of late Archaean oblique convergence. Precambrian Res 99:91–111

Chamberlain KR, Frost CD, Frost BR (2003) Early Archean to Mesoproterozoic evolution of the Wyoming Province: Archean origins to modern lithospheric architecture. Can J Earth Sci 40:1357–1374

Champion DC, Smithies RH (1999) Archaean granites of the Yilgarn and Pilbara cratons: secular changes. In: Barbarin B (ed) The origin of granites and related rocks, fourth Hutton symposium abstracts, 137. BRGM, Clermont-Ferrand

Champion DC, Smithies RH (2007) Geochemistry of paleoarchaean granites of the East Pilbara terrain, Pilbara Craton, Western Australia: implications for early Archaean crustal growth. In: Van Kranendonk MJ, Smithies RH, Bennett VC (eds) Earth's oldest rocks, Developments in Precambrian geology 15. Elsevier, Amsterdam, pp 339–409

Chauvel C, Dupre B, Arndt NT (1993) Pb and Nd isotopic correlation in Belingwe komatiites and basalts. In: Bickle MJ, Nisbet EG (eds) The geology of the Belingwe greenstone belt. Geological Society Zimbabwe special publication 2. A. A. Balkema, Rotterdam, pp 167–174

Cheney S (1996) Sequence stratigraphy and plate tectonic significance of the Transvaal succession of southern Africa and its equivalent in Western Australia. Precambrian Res 79:3–24

Chou CL (1978) Fractionation of siderophile elements in the Earthís upper mantle. In: Proceeding of the lunar and planet science conference 9th. pp 219–230

Chyba CF (1993) The violent environment of the origin of life: progress and uncertainties. Geochim Cosmochim Acta 57:3351–3358

Chyba CF, Sagan C (1996) Comets as the source of prebiotic organic molecules for the early Earth. In: Thomas PJ, Chyba CF, McKay CP (eds) Comets and the origin and evolution of life. Springer, New York, pp 147–174

Cloete M (1999) Aspects of volcanism and metamorphism of the Onverwacht Group lavas in the southwestern portion of the Barberton Greenstone Belt. Memoir Geological Society of South Africa 84, 232 pp

Cloud P (1968) Atmospheric and hydrospheric evolution of the primitive Earth. Earth Sci 160:729–738

Cloud P (1972) Precambrian of North America. Am J Sci 272:6

Cloud P (1973) Paleoecological significance of the banded iron formation. Econ Geol 68:1135–1143

Cohen BA, Swindle TD, Kring DA (2000) Support for the lunar cataclysm hypothesis from lunar meteorite melt ages. Science 290:1754–1756

Coleman RG, Peterman ZE (1975) Oceanic plagiogranites. J Geophys Res 80:1099–1108

Collerson KD, Bridgwater D (1979) Metamorphic development of early Archaean tonalitic and trondhjemitic gneisses: Saglek area, Labrador. In: Barker F (ed) Trondhjemites, dacites and related rocks. Elsevier, Amsterdam, pp 205–274

Collerson KD, Fryer BJ (1978) The role of fluids in the formation and subsequent development of early continental crust. Contrib Mineral Petrol 67:151–169

Compston W, Kröner A (1988) Multiple zircon growth within early Archean Tonalitic gneiss from the ancient Gneiss Complex, Swaziland. Earth Planet Sci Lett 87:13–28

Condie KC (2001) Mantle plumes and their record in earth history. Cambridge University Press, Cambridge

Condie KC, Hunter DR (1976) Trace element geochemistry of Archaean granitic rocks from the Barberton region, south Africa. Earth Planet Sci Lett 29:389–400

Crowley JL, Myers JS, Sylvester PJ, Cox RA (2005) Detrital zircons from the Jack Hills and Mount Narryer, Western Australia: evidence for diverse >4.0 Ga source rocks. J Geol 113:239–263

Culler TS, Becker TA, Muller RA, Renne PR (2000) Lunar impact history from 39Ar/40Ar dating of glass spherules. Science 287:1785–1789

Dalziel IWD (1991) Pacific margins of Laurentia and East Antarctica–Australia as a conjugate rift pair: evidence and implications for an Eocambrian supercontinent. Geology 19:598–601

Dann JC (2000) The Komati Formation, Barberton Greenstone Belt, South Africa, Part I: new map and magmatic architecture. S Afr J Earth Sci 103:47–68

Dann JC, Wilson AH, Cloete M (1998) Field excursion C1: Komatiites in the Barberton and Nondweni greenstone belts. In: IAVCEI – magmatic diversity: volcanoes and their roots. Cape Town, 61 pp

Darwin C (1859) On the origin of species by means of natural selection

Dauphas N, van Zuilen M, Wadhwa M, Davis AM, Marty B, Janney PE (2004) Clues from Fe isotope variations on the origin of early Archean BIFs from Greenland. Science 306:2077–2080

Davies GF (1995) Punctuated tectonic evolution of the Earth. Earth Planet Sci Lett 136:363–380

Davies P (1999) The Fifth Miracle: the search for the origin and meaning of life. Touchstone Book, New York, 293 pp

Davies RD, Allsopp HL (1976) Strontium isotope evidence relating to the evolution of the lower Precambrian crust in Swaziland. Geology 4:553–556

Davis DW (2002) U-Pb geochronology of Archean metasedimentary rocks in the Pontiac and Abitibi subprovinces, Quebec, constraints on timing, provenance and regional tectonics. Precambrian Res 115:97–117

Dawes R, Smithies RH, Centofantji J, Podmore DC (1995) Sunrise Hill unconformity: a newly discovered regional hiatus between Archaean granites and greenstones in the northeastern Pilbara Craton. Aust J Earth Sci 42:635–639

de Ronde CEJ (1991) Structural and geochronological relationships and fluid/rock interaction in the central part of the ~3.2–3.5 Ga Barberton greenstone belt, South Africa. Ph.D. thesis, University of Toronto, 370 pp

de Ronde CEJ, Kamo SL (2000) An Archaean Arc-Arc collisional event: a short-lived (ca. 3Myr) episode, Weltevreden area, Barberton greenstone belt, South Africa. J Afr Earth Sci 30(2):219–248

de Wit MJ (1998) On Archean granites, greenstones, cratons and tectonics: does the evidence demand a verdict? Precambrian Res 91:181–226

de Wit MJ, Armstrong R, Hart RJ, Wilson AH (1987) Felsic igneous rocks within the 3.3 to 3.5 Ga Barberton greenstone belt: high crustal level equivalents of the surrounding tonalite-trondhjemite terrain, emplaced during thrusting. Tectonics 6:529–549

DeLaeter JR, Blockley JG (1972) Granite ages within the Archaean Pilbara Block, Western Australia. J Geol Soc Aust 19:363–370

DeLaeter JR, Lewis JD, Blockley JG (1975) Granite ages within the Shaw batholith, Pilbara region, Western Australia. Geol Surv West Aust Ann Rep 1974:73–79

DePaolo DJ (1983) The mean life of continents: estimates of continent recycling rates from Nd and Hf isotopic data and implications of mantle structure. Geophys Res Lett 10:705–708

Dewey J, Spall H (1976) Pre-Mesozoic plate tectonics how far back in history can the Wilson cycle be extended. Geology 3:422–424

Dickin A (1995) Radiogenic isotope geology. Cambridge University Press, Cambridge, 452 pp

Diener JFA, Stevens G, Kisters AFM, Poujol M (2005) Metamorphism and exhumation of the basal parts of the Barberton greenstone belt, South Africa: constraining the rates of Mesoarchaean tectonism. Precambrian Res 143(1–4):87–112

Drennan GR, Robb LJ, Meyer FM, Armstrong RA, de Bruiyn H (1990) The nature of the Archaean basement in the hinterland of the Witwatersrand Basin: II. A crustal profile west of the Welkom Goldfield and comparisons with the Vredefort crustal profile. South Afr J Geol 93:41–53

Drury SA (1978) REE distribution in high-grade Archaean gneiss complex in Scotland: implications for the genesis of ancient sialic crust. Precambrian Res 7:237–257

Duck LJ, Glikson M, Golding SD, Webb RE (2007) Microbial remains and other carbonaceous forms from the 3.24 Ga Sulphur Springs black smoker deposit, Western Australia. Precambrian Res 154:205–220

Duck LJ, Glikson M, Golding SD, Webb R, Riches J, Baiano J, Sly L (2008) Geochemistry and nature of organic matter in 35 Ga rocks from Western Australia. Geochim Cosmochim Acta 70:1457–1470

Dunlop JSR, Buick R (1981) Archaean epiclastic sediments derived from mafic volcanics, North Pole, Pilbara Block, Western Australia. J Geol Soc Aust 7:225–233

Dunlop JSR, Muir MD, Milne VA, Groves DI (1978) A new microfossil assemblage from the Archaean of Western Australia. Nature 274:676–678

Durney DW (1972) A major unconformity in the Archaean, Jones Creek, Western Australia. J Geol Soc Aust 19:251–259

Dymek RF, Klein C (1988) Chemistry, petrology and origin of banded iron-formation lithologies from the 3800 Ma Isua supracrustal belt, West Greenland. Precambrian Res 37:247–302

Dziggel A, Stevens G, Poujol M, Armstrong RA (2006) Contrasting source components of clastic meta-sedimentary rocks in the lowermost formations of the Barberton greenstone belt. In: Reimold WU, Gibson GL (eds) Processes on the early Earth. Geological Society of America special papers 405. Geological Society of America, Boulder, pp 157–172

Eglington BM, Armstrong RA (2004) The Kaapvaal Craton and adjacent orogens, southern Africa: a geochronological database and overview of the geological development of the craton. S Afr J Geol 107:13–32

Eigenbrode JL, Freeman KH (2006) Late Archaean rise of aerobic microbial ecosystems. Proc Natl Acad Sci U S A 103:15759–15764

Eriksson KA (1978) Alluvial and destructive beach facies from the Archaean Moodies Group, Barberton Mountain Land, South Africa and Swaziland. In: Miall AD (ed) Fluvial sedimentology. Canadian Society of Petroleum Geologists Memoir 5. Canadian Society of Petroleum Geologists, Calgary, pp 287–311

Eriksson KA (1979) Marginal marine depositional processes from the Archaean Moodies Group, Barberton Mountain Land, South Africa: evidence and significance. Precambrian Res 8:153–182

Eriksson KA (1980) Transitional sedimentation styles in the Moodies and Fig Tree Groups, Barberton Mountain Land, South Africa: evidence favoring an Archaean continental margin. Precambrian Res 12:141–160

Eriksson KA (1981) Archaean platform-trough sedimentation, East Pilbara Block, Australia. In: Glover JE, Groves DI (eds) Archaean geology, second international Archaean symposium Perth 1980. Geological Society of Australia special publication 7, Perth. Geological Society of Australia, Sydney, pp 235–244

Eriksson KA (1982) Geometry and internal characteristics of Archaean submarine channel deposits, Pilbara Block, Western Australia. J Sed Petrol 52:383–393

Eriksson KA, Krapez B, Fralick W (1994) Sedimentology of Archean greenstone belts: signatures of tectonic evolution. Earth Sci Rev 37:1–88

Ermanovics IF, Davidson WL (1976) The Pikwitone granulites in relation to the northwest Superior Province in the Canadian Shield. In: Windley BF (ed) Early history of the earth. Wiley, London, pp 331–350

Ernst RE, Buchan KL, Prokoph A (2004) Large Igneous Province record through time. In: Eriksson PG et al (eds) The Precambrian Earth: tempos and events in Precambrian time. Develop Precamb Geol 12. Elsevier, Amsterdam, pp 173–180

Farquhar J, Bao H, Thiemens M (2000) Atmospheric influence of Earth's earliest sulfur cycle. Science 289:756–758

Farrow DJ, Harmer RE, Hunter DR, Eglington BM (1990) Rb-Sr and Pb-Pb dating of the Anhalt leucotonalite, northern Natal. South Afr J Geol 93:696–701

Farquhar J, Peters M, Johnston DT, Strauss H, Masterson A, Wiechert U, Kaufman AJ (2007) Isotopic evidence for Mesoarchaea anoxia and changing atmospheric sulphur chemistry. Nature 449:706–709

Fedo CM, Eriksson KA, Krogstad EJ (1996) Geochemistry of shales from Archaean (~3.0 Ga) Buhwa Greenstone Belt, Zimbabwe: implications for provenance and source area weathering. Geochim Cosmochim Acta 60:1751–1763

Ferguson J, Currie KL (1972) Silicate immiscibility in the ancient "Basalts" of the Barberton Mountain Land, Transvaal. Nat Phys Sci 235:86–89

Fitzsimons ICW (2000) Grenville-age basement provinces in East Antarctica: evidence for three separate collisional orogens. Geology 28:879–882

Flament N, Coltice N, Rey P (2013) The evolution of the 87Sr/86Sr of marine carbonates does not constrain continental growth. Precambrian Res 229:177–188

Fletcher JA (2003) The Geology, geochemistry and fluid inclusion characteristics of the Wyldsdale Gold-Bearing Pluton, Northwest Swaziland. University of the Witwatersrand, Johannesburg, 215 pp

Fletcher IR, Rosman KJR, Williams IR, Hickman AH, Baxter JL (1984) Sm-Nd geochronology of greenstone belts in the Yilgarn Block, Western Australia. Precambrian Res 26:333–361

Floyd PA, Winchester JA (1975) Magma type and tectonic setting discrimination using immobile elements. Earth Planet Sci Lett 27:211–218

Folinsbee RE, Baadsgaard H, Cumming GL, Green DC (1968) A very ancient island arc. In: Knopoff et al (eds) The crust and upper mantle of the Pacific Area. American Geophysical Union Monograph 12. American Geophysical Union, Washington, pp 441–448

French BM (1998) Traces of catastrophe – a handbook of shock metamorphic effects in terrestrial meteorite impact structures. Lunar Planet Sci Inst Contrib 954:120

Frey R, Rosing MT (2005) Search for traces of the late heavy bombardment on Earth – results from high precision chromium isotopes. Earth Planet Sci Lett 236:28–40

Frey R, Polat A, Meibom A (2004) The Hadean upper mantle conundrum; evidence for source depletion and enrichment from Sm-Nd, Re-Os, and Pb isotopic compositions in 3.71 Ga boninite-like metabasalts from the Isua supracrustal belt, Greenland. Geochim Cosmochim Acta 68:1645–1660

Friend CRL, Hughs DJ (1977) Archaean aluminous ultrabasic rocks with primary igneous textures from the Fiskenaesset region, southern west Greenland. Earth Planet Sci Lett 36:157–167

Friend CRL, Bennett VC, Nutman AP, Norman M (2007) Seawater-like trace element signatures (REE + Y) of Eoarchaean chemical sedimentary rocks from southern West Greenland, and their corruption during high-grade metamorphism. Contrib Miner Petrol 183(4):725–737

Frost CD, Frost BR (1993) The Archean history of the Wyoming province. In: Snoke AW, Steidtmann JR, Roberts SM (eds) Geology of Wyoming. Geological Survey of Wyoming Memoir, 5. Geological Survey of Wyoming Memoir, Laramie, pp 58–77

Frost BR, Frost CD, Cornia ME, Chamberlain KR, Kirkwood R (2006) The Teton-Wind River domain: a 2.68–2.67 Ga active margin in the western Wyoming Province. Can J Earth Sci 43:1489–1510

Froude CF, Ireland TR, Kinny PD, Williams IS, Compston W, Williams IR, Myers JS (1983) Ion-microprobe identification of 4100–4200 Myr old terrestrial zircons. Nature 304:616–618

Fryer BJ (1977) Rare earth evidence in iron-formations for changing Precambrian oxidation states. Geochim Cosmochim Acta 41:361–367

Fryer BJ, Fyfe WS, Kerrich R (1979) Archaean volcanogenic oceans. Chem Geol 24:25–33

Garde AA, McDonald I, Dyck B, Keulen N (2012) Searching for giant ancient impact structures on Earth: the Meso-Archaean Maniitsoq structure, West Greenland. Earth Planet Sci Lett 2012:337–338

Garrels RM, Perry EA, MacKenzie FT (1973) Genesis of Precambrian iron formations and the development of atmospheric oxygen. Econ Geol 68:1173–1179

Gee RD, Baxter JL, Wilde SA, Williams IR (1981) Crustal development in the Archaean Yilgarn Block, Western Australia. In: Glover JA, Groves DI (eds) Archaean geology. Geological Society of Australia special publications 7. Geological Society of Australia, Sydney, pp 43–56

Genest S, Robert F, Goulet N (2010) Otish Nasin: discovery of nearly 2.1 Ga shocked rocks potentially owning to a D>500 km impact structure, Quebec, Canada. Abstract 74th Annual Meeting Meteorological Society A76

Glass BP, Burns CA (1988) Microkrystites: a new term for impact-produced glassy spherules containing primary crystallites. Proc Lunar Planet Sci Conf 18:455–458

Glikson AY (1970) Geosynclinal evolution and geochemical affinities of early Precambrian systems. Tectonophysics 9:397–433

Glikson AY (1971) Primitive Archaean element distribution patterns: chemical evidence and tectonic significance. Earth Sci Planet Lett 12:309–320

Glikson AY (1972a) Petrology and geochemistry of metamorphosed Archaean ophiolites, Kalgoorlie-Coolgardie, Western Australia. Aust Bur Miner Resour Bull 125:121–189

Glikson AY (1972b) Early Precambrian evidence of a primitive ocean crust and island arc nuclei of sodic granite. Bull Geol Soc Am 83:3323–3344

Glikson AY (1976a) Trace element geochemistry and origin of early Precambrian acid igneous series, Barberton Mountain Land, Transvaal. Geochim Cosmochim Acta 40:1261–1280

Glikson AY (1976b) Stratigraphy and evolution of primary and secondary greenstones: significance of data from Shields of the southern hemisphere. In: Windley BF (ed) The early history of the earth. Wiley, London, pp 257–277

Glikson AY (1976c) Archaean to early Proterozoic shield elements: relevance of plate tectonics. Geol Assoc Can Spec Publ 14:489–516

Glikson AY (1978) Archaean granite series and the early crust, Kalgoorlie System, Western Australia. In: Windley BF, Naqvi SM (eds) Archaean geochemistry. Elsevier, Amsterdam, pp 151–174

Glikson AY (1979a) Early Precambrian tonalite-trondhjemite sialic nuclei. Earth Sci Rev 15:1–73

Glikson AY (1979b) On the foundation of the Sargur Group. J Geol Soc India 20:248–255

Glikson AY (1980) Uniformitarian assumptions, plate tectonics and the Precambrian Earth. In: Kroner A (ed) Precambrian plate tectonics. Elsevier, Amsterdam, pp 91–104

Glikson AY (1982) The early Precambrian crust with reference to the Indian Shield – an essay. Geol Soc India J 23:581–603

Glikson AY (1983) Geochemistry of Archaean tholeiitic basalt and high-Mg to peridotitic komatiite suites, with petrogenetic implications. Geol Soc India Mem 4:183–219

Glikson AY (1995) The Giles Mafic-ultramafic Complex and Environs, Western Musgrave Block, Central Australia. Aust Geol Surv Organ J Aust Geol Geophys 16:11–12, 193 pp

Glikson AY (2001) The astronomical connection of terrestrial evolution crustal effects of post-3.8 Ga mega-impact clusters and evidence for major 3.2 Ga bombardment of the Earth–Moon system. J Geodyn 32:205–229

Glikson AY (2004a) Early Precambrian asteroid impact-triggered tsunami: excavated seabed debris flows exotic boulders and turbulence features associated with 3.47–2.47 Ga-old asteroid impact fallout units, Pilbara Craton, Western Australia. Astrobiology 4:1–32

Glikson AY (2004b) A map of asteroid impact signatures of the Pilbara Craton. Geological survey of Western Australia report 102 Plate 1

Glikson AY (2005) Geochemical and isotopic signatures of Archaean to early Proterozoic extraterrestrial impact ejecta/fallout units. Aust J Earth Sci 52:785–799

Glikson AY (2006) Asteroid impact ejecta units overlain by iron rich sediments in 3.5–2.4 Ga terrains Pilbara and Kaapvaal cratons: accidental or cause–effect relationships? Earth Planet Sci Lett 246:149–160

Glikson AY (2007) Siderophile element patterns, PGE nuggets and vapour condensation effects in Ni-rich quench chromite-bearing microkrystite spherules, 3.24 Ga S3 impact unit, Barberton greenstone belt, Kaapvaal Craton, South Africa. Earth Planet Sci Lett 253:1–16

Glikson AY (2008) Field evidence of Eros-scale asteroids and impact-forcing of Precambrian geodynamic episodes, Kaapvaal (South Africa) and Pilbara (Western Australia) Cratons. Earth Planet Sci Lett 267:558–570

Glikson AY (2010) Archaean asteroid impacts, banded iron formations and MIF-S anomalies: a discussion. Icarus 207:39–44

Glikson AY (2013) The asteroid impact connection of planetary evolution. Springer-Briefs, Dordrecht, 150 pp

Glikson AY, Allen C (2004) Iridium anomalies and fractionated siderophile element patterns in impact ejecta, Brockman Iron Formation, Hamersley Basin, Western Australia: evidence for a major asteroid impact in simatic crustal regions of the early Proterozoic earth. Earth Planet Sci Lett 20:247–264

Glikson AY, Hickman AH (1981) Geochemical stratigraphy of Archaean mafic-ultramafic volcanic successions, eastern Pilbara Block, Western Australia. In: Glover JE, Groves DI (eds) Archaean geology. Geological Society of Australia special publications 7. Geological Society of Australia, Sydney, pp 287–300

Glikson AY, Jahn B (1985) REE and LIL elements, eastern Kaapvaal shield, south Africa: evidence of crustal evolution by 3-stage melting. Geol Soc Can Spec Pap 28:303–324

Glikson AY, Lambert IB (1976) Vertical zonation and petrogenesis of the early Precambrian crust in Western Australia. Tectonophysics 30:55–89

Glikson AY, Sheraton JW (1972) Early Precambrian trondhjemitic suites in Western Australia and northwestern Scotland and the geochemical evolution of shields. Earth Planet Sci Lett 17:227–242

Glikson AY, Vickers J (2006) The 3.26–3.24 Ga Barberton asteroid impact cluster: tests of tectonic and magmatic consequences, Pilbara Craton, Western Australia. Earth Planet Sci Lett 241:11–20

Glikson AY, Vickers J (2007) Asteroid mega-impacts and Precambrian banded iron formations: 2.63 Ga and 2.56 Ga impact ejecta/fallout at the base of BIF/argillite units, Hamersley Basin, Pilbara Craton, Western Australia. Earth Planet Sci Lett 254:214–226

Glikson AY, Vickers J (2010) Asteroid impact connections of crustal evolution. Aust J Earth Sci 57:79–95

Glikson AY, Davy R, Hickman AH, Pride C, Jahn B (1987) Trace elements geochemistry and petrogenesis of Archaean felsic igneous units, Pilbara Block, Western Australia. Australian Bureau of Mineral Resources Record 87/30

Glikson AY, Davy R, Hickman AH (1989) Trace metal distribution in basalts, Pilbara craton, Western Australia, with stratigraphic-geochemical implications. BMR Record 1991/46

Glikson AY, Allen C, Vickers J (2004) Multiple 3.47-Ga-old asteroid impact fallout units, Pilbara Craton, Western Australia. Earth Planet Sci Lett 221:383–396

Glikson M, Duck LJ, Golding SD, Hofmann A, Bolhar R, Webb R, Baiano JCF, Sly LI (2008) Microbial remains in some earliest Earth rocks: comparison with a potential modern analogue. Precambrian Res 164:187–200

Glikson AY, Uysal IT, Fitz Gerald JD, Saygin E (2013) Geophysical anomalies and quartz microstructures, Eastern Warburton Basin, North-east South Australia: tectonic or impact shock metamorphic origin? Tectonophysics 589:57–76

Goderis S, Simonson BM, McDonald I, Hassler SW, Izmer A, Belza J, Terryn H, VanHaeck F, Claeys P (2013) Ni-rich spinels and platinum group element nuggets condensed from a Late Archaean impact vapour cloud. Earth Planet Sci Lett 376:87–98

Gold T (1999) The deep hot biosphere. Springer, New York, 235 pp

Goldich SS, Hedge CE (1974) 3,800-Myr granitic gneisses in south-western Minnesota. Nature 252:467–468

References

Golding SD, Duck LJ, Young E, Baublys KA, Glikson M (2011) Earliest seafloor hydrothermal systems on earth: comparison with modern analogues. In: Golding S, Glikson MV (eds) Earliest life on earth: habitats, environments and methods of detection. Springer, Dordrecht, pp 1–15

Goodwin AM (1968) Archean protocontinental growth and early crustal history of the Canadian Shield. In: 23rd International geological congress, Prague 1. pp 69–89

Goodwin AM (1981) Precambrian perspectives. Science 213:55–61

Goodwin AM, Monster J, Thode HG (1976) Carbon and sulfur isotope abundances in Archaean iron-formations and early Precambrian life. Econ Geol 71:870–891

Gorbatschev R, Bogdanova S (1993) Frontiers in the Baltic Shield. Precambrian Res 64:3–21

Gorman AR, Clowes RM, Ellis RM, Henstock TJ, Spence GD, Keller GR, Levander A, Snelson CM, Burianyk MJA, Kanasewich ER, Asuden I, Hajnal Z, Miller KC (2002) Deep Probe: imaging the roots of western North America. Can J Earth Sci 39:375–398

Gould SJ (1990) Wonderful life: the Burgess Shale and the nature of history. Norton 347 pp

Graf JL (1978) Rare earth elements, iron formations and sea water. Geochim Cosmochim Acta 42:1845–1850

Green DH (1972) Archaean greenstone belts may include equivalents of lunar maria? Earth Planet Sci Lett 15:263–270

Green DH (1981) Petrogenesis of Archaean ultramafic magmas and implications for Archaean tectonics. In: Kroner A (ed) Precambrian plate tectonics. Elsevier, Amsterdam, pp 469–489

Green DC, Baadsgaard H (1971) Temporal evolution and petrogenesis of an Archaean crustal segment at Yellowknife, NWT. Can J Petrol 12:177–217

Green DH, Ringwood AE (1967) An experimental investigation of the gabbro to eclogite transformation and its petrological applications. Geochim Cosmochim Acta 31:767–833

Green TH, Ringwood AE (1977) Genesis of the calc-alkaline igneous rock suite. Contrib Mineral Petrol 18:105–162

Grieve RAF (2006) Impact structures in Canada. Geological Association of Canada, St. John, 210 pp

Grieve RAF, Dence MR (1979) The terrestrial cratering record: II the crater production rate. Icarus 38:230–242

Griffin TJ (1990) North Pilbara granite-greenstone terrain. Geol Surv West Aust Mem 3:128–158

Griffin WL, O'Reilly SY, Abe N, Aulbach S, Davies RM, Pearson NJ, Doyle BJ, Kivi K (2003) The origin and evolution of Archean lithospheric mantle. Precambrian Res 127:19–41

Griffin WL, Belousova EA, Shee SR, Pearson NJ, O'Reilly SYO (2004) Archaean crustal evolution in the northern Yilgarn Craton: U-Pb and Hf-isotope evidence from detrital zircons. Precambrian Res 131:231–282

Griffin WL, Belousova EA, O'Neill C, O'Reilly SY, Malkovets V, Pearson NJ, Spetsius S, Wilde SA (2014) The world turns over: Hadean–Archean crust–mantle evolution. Lithos 189:2–15

Groves DI, Batt WD (1984) Spatial and temporal variations of Archaean metallogenic associations in terms of evolution of granitoid-greenstone terrains with particular emphasis on the Western Australian Shield. In: Kroner A et al (eds) Archaean geochemistry. Springer, Berlin, pp 73–98

Gruau G, Jahn B, Glikson AY, Davy R, Hickman CC (1987) Age of the Archaean Talga-Talga Subgroup, Pilbara Block, Western Australia, and early evolution of the mantle: new Sm-Nd evidence. Earth Planet Sci Lett 85:105–116

Hallberg JA (1971) Geochemistry of Archaean Volcanic Belts in the Eastern Goldfields Region of Western Australia. J Petrol 13:45–56

Hallberg JA, Glikson AY (1980) Archaean granite- greenstone terrains of Western Australia. In: Hunter DR (ed) Precambrian of the southern continents. Elsevier, Amsterdam, pp 33–103

Halverson GP, Hofmann PF, Schrag DP, Maloof AC, Adam C, Hugh A, Rice N (2005) Toward a Neoproterozoic composite carbon-isotope record. Geol Soc Am Bull 117:1181–1207

Hamilton WB (2003) An alternative Earth. Geol Soc Am Today 13:412

Hamilton PJ, Evensen NM, O'Nions RK, Glikson AY, Hickman AH (1981) Sm-Nd dating of the Talga-Talga Subgroup, Warrawoona Group, Pilbara Block, Western Australia. In: Glover JE, Groves DI (eds) Archaean Geology. Geological Society of Australia special publications 7. Geological Society of Australia, Sydney, pp 187–192

Hanson GN (1975) Geochemistry and origin of the early Precambrian crust of northeastern Minnesota. Geochim Cosmochim Acta 39:325–362

Harley SL (2003) Archaean to Pan-African crustal development and assembly of East Antarctica: metamorphic characteristics and tectonic implications. In: Yoshida M, Windley BF (eds) Proterozoic East Gondwana: supercontinent assembly and breakup. Geological Society of London special publications 206. Geological Society, London, pp 203–230

Harley SL, Kelly NM (2007) Ancient Antarctica: the Archaean of the East Antarctic shield. In: Van Kranendonk MJ, Smithies RH, Bennett VC (eds) Earth's oldest rocks. Developments in Precambrian geology 15. Elsevier, Amsterdam, pp 149–186

Harrison TM, Blichert-Toft J, Müller W, Albarede F, Holden P, Mojzsis SJ (2005) Heterogeneous Hadean hafnium: evidence of continental crust at 4.4 to 4.5 Ga. Science 310:1947–1950

Hart R, Moser D, Andreoli M (1999) Archaean age for the granulite facies metamorphism near the center of the Vredefort structure, South Africa. Geology 27(12):1091–1094

Hartmann WK, Ryder G, Dones L, Grinspoon DH (2000) The time-dependent intense bombardment of the primordial Earth-Moon system. In: Canup R, Righter K (eds) Origin of the Earth and Moon. University of Arizona Press, Tucson, pp 493–512

Hassler SW (1993) Depositional history of the Main Tuff Interval of the Wittenoom Formation, late Archaean-early Proterozoic Hamersley Group, Western Australia. Precambrian Res 60:337–359

Hassler SW, Simonson BM (2001) The sedimentary record of extraterrestrial impacts in deep shelf environments evidence from the early Precambrian. J Geol 109:1–19

Hassler SW, Robey HF, Simonson BM (2000) Bedforms produced by impact-generated tsunami, ~2.6 Ga Hamersley basin, Western Australia. Sediment Geol 135:283–294

Hassler SW, Simonson BM, Sumner DY, Murphy M (2005) Neoarchean impact spherule layers in the Fortescue and Hamersley group, Western Australia: stratigraphic and depositional implications of re-correlation. Aust J Earth Sci 52:759–771

Hassler SW, Simonson BM, Sumner DY, Bodin L (2011) Paraburdoo spherule layer (Hamersley Basin, Western Australia): distal ejecta from a fourth large impact near the Archaean-Proterozoic boundary. Geology 39:307–310

Head JW (1976) Lunar volcanism in space and time. Rev Geophys Space Phys 14:265–300

Heinrichs TK, Reimer TO (1977) A sedimentary barite deposit from the Archean Fig Tree Group of the Barberton Mountain Land (South Africa). Econ Geol 72:1426–1441

Heubeck C, Lowe DR (1994) Depositional and tectonic setting of the Archean Moodies Group, Barberton Greenstone Belt, South Africa. Precambrian Res 68:257–290

Heubeck C, Lowe DR (1999) Sedimentary petrography and provenance of the Archean Moodies Group, Barberton Greenstone Belt. In: Lowe DR, Byerly GR (eds) Geologic evolution of the Barberton Greenstone Belt, South Africa. Geological Society of America special paper 329. Geological Society of America, Boulder, pp 259–286

Hickman AH (1975) Precambrian structural geology of part of the Pilbara region. In: Annual Report 1974. Western Australia Geological Survey, pp 68–73

Hickman AN (1977) New and revised definitions of rock units in the Warrawoona Group, Pilbara Block, Western Australia. Geol Surv West Aust Ann Rep 1976:58

Hickman AH (1981) Crustal evolution of the Pilbara Block. In: Glover JE, Groves DI (eds) Archaean geology: second international Archaean symposium, Perth, 1980. Geological Society of Australia special publications 7. Geological Society of Australia, Sydney, pp 57–69

Hickman AH (1983) Geology of the Pilbara Block and its environs. West Australia Geological Survey Bulletin 127. Geological Survey of Western Australia, Perth, 268 pp

Hickman AH (1990) Geology of the Pilbara Craton. In: Ho SE, Glover JE, Myers JS, Muhling JR (eds) Third international Archaean symposium, excursion guidebook. University of Western Australia, Nedlands, pp 1–13

Hickman AH (2004) Two contrasting granite–greenstones terrains in the Pilbara Craton, Australia: evidence for vertical and horizontal tectonic regimes prior to 2900 Ma. Precambrian Res 131:153–172

Hickman AH (2012) Review of the Pilbara Craton and Fortescue Basin, Western Australia: crustal evolution providing environments for early life. Island Arc 21:1–31

Hickman AH, Lipple L (1978) Marble Bar, Western Australia, Sheet 50-8. Geological Survey of Western Australia 1:250 000 Geological Series Explanatory Notes, Perth

Hickman AH, Van Kranendonk MJ (2004) Diapiric processes in the formation of Archaean continental crust, East Pilbara Granite-Greenstone Terrain, Australia. In: Eriksson PG et al (eds) The Precambrian Earth: tempos and events in Precambrian. Time, developments in Precambrian geology 12. Elsevier, Amsterdam, pp 118–139

Hickman AH, Van Kranendonk MJ (2008) Archean crustal evolution and mineralization of the northern Pilbara Craton – a field guide. Geol Surv West Aust Rec 2008(13):1–79

Hickman AH, Smithies RH, Tyler IM (2010) Evolution of active plate margins: West Pilbara Superterrain, De Grey Superbasin, and Fortescue and Hamersley Basins – a field guide. Geol Surv West Aust Rec 2010(3):1–74

Hietanen A (1975) Generation of potassium-poor magmas in the northern Sierra Nevada and the Sveconfennian of Finland. U S Geol Surv J Res 3:631–645

Hill RM (1997) Stratigraphy, structure and alteration of hanging wall sedimentary rocks at the Sulphur Springs volcanogenic massive sulphide (VMS) prospect, East Pilbara Craton, Western Australia. B.Sc Hon. thesis, University of Western Australia

Hirner A (2001) Geology and gold mineralization in the Madibe Greenstone Belt, Eastern part of the Kraaipan Terrain, Kaapvaal Craton, South Africa. University of the Witwatersrand, Johannesburg, 220 pp

Hodges KV, Bowring SA, Coleman DS, Hawkins DP, Davidek KL (1995) Multi-stage thermal history of the ca. 4.0 Ga Acasta gneisses. In: American Geophysical Union fall meeting. p F708

Hofmann AW (1988) Chemical differentiation of the Earth: the relationship between mantle, continental crust, and oceanic crust. Earth Planet Sci Lett 90:297–314

Hofmann PF, Schrag DP (2000) Snowball earth. Sci Am 282:68–75

Hofmann PF, Kaufman AJ, Halverson GP, Schrag DP (1998) A Neoproterozoic snowball Earth. Science 281:1342–1346

Hofmann HJ, Grey K, Hickman AH, Thorpe RI (1999) Origin of 3.45 Ga Coniform Stromatolites in the Warrawoona Group, Western Australia. Bull Geol Soc Am 111:1256–1262

Holland HD (1984) The chemical evolution of the atmospheres and oceans. Princeton University Press, Princeton, 582 pp

Holland HD (1994) Early Proterozoic atmosphere change. In: Bengston S (ed) Nobel symposium 84, early life on Earth. Columbia University Press, New York, pp 237–244

Holland HD (2005) Sedimentary mineral deposits and the evolution of Earth's near surface environments. Econ Geol 100:1489–1500

Holmes A (1965) Principles of physical geology. Ronald Press, New York, 1288 pp

Horstwood MSA, Nesbitt RW, Noble SR, Wilson JF (1999) U-Pb zircon evidence for an extensive early Archean craton in Zimbabwe: a reassessment of the timing of craton formation, stabilization, and growth. Geology 27:707–710

Hui SM, Norman MD, Jourdan F (2009) Tracking formation and transport of Apollo 16 lunar impact glasses through chemistry and dating. 9th Australian space science conference, Sydney, Australia, 28–30 Sept 2009

Hunter DR (1974) Crustal development in the Kaapval craton: part 1 – the Archaean. Precambrian Res 1:259–294

Hunter DR, Barker F, Millard HT (1978) The geochemical nature of the Archacan ancient gneiss complex and granodiorite suite, Swaziland: a preliminary study. Precambrian Res 7:105–127

Hunter DR, Smith RG, Sleigh DWW (1992) Geochemical studies of Archaean granitoid rocks in the southeastern Kaapvaal Province: implications for crustal development. J Afr Earth Sci 15(1):127–151

Hurley PM, Hughs H, Faure G, Fairbairn HW, Pinson WH (1962) Radiogenic strontium 87 model of continent formation. J Geophys Res 67:5315–5334

Hurst RW (1978) Sr evolution in west Greenland–Labrador craton: a model for early Rb depletion in the mantle. Geochim Cosmochim Acta 42:39–44

Hurst J, Krapež B, Hawke P (2013) Stratigraphy of the Marra Mamba iron formation within the Chichester Range and its implications for iron ore genesis at Roy Hill – evidence from deep diamond drill holes within the East Fortescue Valley. Iron Ore, Australian Institute of Mining Metal paper 95

Hutton J (1788) The theory of the Earth. Trans Geol Soc Edinb 1:209–304

Iizuka T, Komiya T, Ueno Y, Katayama I, Uehara Y, Maruyama S, Hirata T, Johnson SP, Dunkley D (2007) Geology and zircon geochronology of the Acasta Gneiss Complex, northwestern Canada: new constraints on its tectono-thermal history. Precambrian Res 153:179–208

Izett GA (1990) The Cretaceous–Tertiary boundary interval, Raton Basin, Colorado and New Mexico, and its content of shock-metamorphosed minerals; evidence relevant to the K/T boundary impact-extinction theory. Geol Soc Am Spec Pap 249:100

Jacobsen SB, Pimentel-Close M (1988) A Nd isotopic study of the Hamersley and Michipicoten banded iron formations: the source of REE and F in Archaean oceans. Earth Planet Sci Lett 87:29–44

Jahn B, Vidal P, Tilton G (1979) Archaean mantle heterogeneity: evidence from chemical and isotopic abundances in Archaean igneous rocks. Philos Trans R Soc Lond A297:353–364

Jahn BM, Glikson AY, Peucat JJ, Hickman AH (1981) REE geochemistry and geochronology of Archaean silicic volcanics and granitoids from the Pilbara Block, Western Australia. Geochim Cosmochim Acta 45:1633–1652

Jahn BM, Gruau G, Glikson AY (1982) Onverwacht Group komatiites, South Africa: REE geochemistry, Sm-Nd age and mantle evolution. Contrib Mineral Petrol 80:25–40

Janardhan AS, Srikantappa C, Ramachandra HM (1978) The Sargur schist complex – an Archaean high grade terrain in south India. In: Windley BF, Naqvi SM (eds) Archaean geochemistry. Elsevier, Amsterdam, pp 127–150

Jelsma HA, Dirks PHGM (2000) Tectonic evolution of a greenstone sequence in northern Zimbabwe: sequential early stacking and pluton diapirism. Tectonics 19:135–152

Jelsma H, Vinyu M, Valbracht P, Davies G, Wijbrans J, Verdurmen E (1996) Constraints on Archean crustal evolution of the Zimbabwe craton: a U-Pb zircon, Sm-Nd and Pb-Pb whole rock isotope study. Contrib Mineral Petrol 124:55–70

Jenner FJ, Bennett VC, Nutman AP (2006) 3.8 Ga arc-related basalts from Southwest Greenland. Geochim Cosmochim Acta 70:A291

Jenner FE, Bennett VC, Yaxley G, Friend CRL, Nebel O (2013) Eoarchean within-plate basalts from southwest Greenland. Geology 41:327–330

Jorgensen UG, Apel PWU, Hatasukawa Y, Frei R, Oshima M, Toh Y, Kimura A (1999) The Earth-Moon system during the Late Heavy Bombardment period. Earth Planet Astrophys (astro-ph. EP) arXiv:0907.4104

Kamber B (2007) The enigma of the terrestrial protocrust: evidence for its former existence and the importance of its complete disappearance. In: Condie KC (ed) Developments in Precambrian geology, vol 15. Elsevier, Amsterdam, pp 75–89

Kamber BS, Collerson KD, Moorbath S, Whitehouse MJ (2003) Inheritance of early Archaean Pb-isotope variability from long-lived Hadean protocrust. Contrib Mineral Petrol 145:25–46

Kamo SL, Davis DW (1994) Reassessment of Archean crust development in the Barberton Mountain Land, South Africa, based on U-Pb dating. Tectonics 13:167–192

Kamo SL, Davis DW, De Wit MJ (1990) U-Pb geochronology of Archean plutonism in the Barberton region, S. Africa: 800 Ma of crustal evolution. In: 7th international geocongress on geochemistry, Canberra, p 53

Kamo SL, Reimold WU, Krogh TE, Colliston WP (1996) A 2.023 Ga age for the Vredefort impact event and a first report of shock metamorphosed zircons in pseudotachylitic breccias and granophyre. Earth Planet Sci Lett 144:369–387

Kasting JF, Ono S (2006) Palaeoclimates: the first two billion years. Philos Trans R Soc Biol Sci 361:917–929

Katz MB (1972) Paired metamorphic belts of the Gondwana Precambrian and plate tectonics. Nature 239:271–273

References

Kerrich R, Polat A (2006) Archean greenstone-tonalite duality: thermochemical mantle convection models or plate tectonics in the early Earth global dynamics? Tectonophysics 415:141–165

Kerrich R, Polat A, Wyman DA, Hollings P (1999) Trace element systematics of Mg to Fe tholeiitic basalt suites of the Superior province: implications for Archean mantle reservoirs and greenstone belt genesis. Lithos 46:163–187

Key RM, Litherland M, Hepworth JV (1976) The evolution of the Archaean crust of NE Botswana. Precambrian Res 3:375–413

Kinny PD, Mass R (2003) Lu-Hf and Sm-Nd isotope systems in zircon. Rev Mineral Geochem 53:327–341

Kinny PD, Wijbrans JR, Froude DO, Williams IS, Compston W (1990) Age constraints on the geological evolution of the Narryer Gneiss Complex, Western Australia. Aust J Earth Sci 37:51–69

Kirschvink JL (1992) In: Schopf JW, Klein C (eds) The proterozoic biosphere. Cambridge University Press, New York, pp 51–52

Kisters AFM, Anhaeusser CR (1995) Emplacement features of Archaean TTG plutons along the southern margin of the Barberton greenstone belt, South Africa. Precambrian Res 75:1–15

Kisters AFM, Belcher RW, Poujol M, Dziggel A (2010) Continental growth and convergence-related arc magmatism in the Mesoarchaean: evidence from the Barberton granitoid-greenstone terrain, South Africa. Precambrian Res 178:15–26

Kisters A, Belcher R, Poujol M, Stevens G, Moyen JF (2006) A 3.2 Ga magmatic arc preserving 50 Ma of crustal convergence in the Barberton terrain, South Africa. In: AGU fall meeting, San Francisco, USA, 11–15 Dec 2006

Kitajima K, Maruyama S, Utsonomita S, Liou JG (2001) Seafloor hydrothermal alteration at an Archaean mid-ocean ridge. J Metamorph Geol 19:581–597

Kiyokawa S, Taira A, Byrne T, Bowring S, Sano Y (2002) Structural evolution of the middle Archaean coastal Pilbara terrain, Western Australia. Tectonics 21:1044

Knauth LP (2005) Temperature and salinity history of the Precambrian ocean: implications for the course of microbial evolution. Palaeogeogr Palaeoclimatol Palaeoecol 219:53–69

Knauth LP, Lowe R (1978) Oxygen isotope geochemistry of cherts from the Onverwacht Group (3.4 billion years) Transvaal, South Africa, with implications for secular variations in the isotopic composition of cherts. Earth Planet Sci Lett 41:209–222

Knauth LP, Lowe DR (2003) High Archean climatic temperatures inferred from oxygen isotope geochemistry of cherts in the 3.5 Ga Swaziland Supergroup, South Africa. Bull Geol Soc Am 115:566–580

Knoll AH, Javaux EJ, Hewitt D, Cohen P (2006) Eukaryotic organisms in Proterozoic oceans. Phil Trans R Soc Lond B 361:1023–1038

Koeberl C (2006) Impact processes on the early Earth. Elements 2:211–216

Kohler EA, Anhaeusser CR, Isachsen C (1993) The bien venue formation: a proposed new unit in the northeastern sector of the Barberton greenstone belt. In: 16th international colloquium on Africa geology, Ezulwini, Swaziland. Swaziland G.S.M.D., pp 186–188

Kojan CJ, Hickman AH (1998) Late Archaean volcanism in the Kylena and Maddina Formations, Fortescue Group, west Pilbara. Geol Surv West Aust Annu Rev 1997–98:43–53

Komiya T, Maruyama S, Masuda T, Nohda S, Hayashi M, Okamoto K (1999) Plate tectonics at 3.8–3.7 Ga: field evidence from the Isua accretionary complex, southern West Greenland. J Geol 107:515–554

Konhausser K, Hamada T, Raiswell R, Morris R, Ferris F, Southam G, Canfield D (2002) Could bacteria have formed the Precambrian banded iron-formations? Geology 30:1079–1082

Kopp RE, Kirschvink JL, Hilburn IA, Nash CZ (2005) The Paleoproterozoic snowball Earth: a climate disaster triggered by the evolution of oxygenic photosynthesis. Proc Natl Acad Sci U S A 102:11131–11136

Krapez B, Barley MB (2008) Late Archaean synorogenic basins of the Eastern Goldfields Superterrain, Yilgarn Craton, Western Australia Part III. Signatures of tectonic escape in an arc-continent collision zone. Precambrian Res 161:183–199

Krapez B, Eisenlohr B (1998) Tectonic settings of Archaean (3325–2775 Ma) crustal-supracrustal belts in the West Pilbara Block. Precambrian Res 88:173–205

Krauskopf KB (1970) A tell of ten plutons. In: Cloud P (ed) Adventures in Earth history. Freeman, San Francisco, pp 54–70

Kroner A (1981) Precambrian plate tectonics. In: Kroner A (ed) Precambmrian Plate Tectonics. Elsevier, Amsterdam, pp 57–90

Kroner A (1991) Tectonic evolution in the Archaean and Proterozoic. Tectonophysics 187:393–410

Kröner A, Tegtmeyer A (1994) Gneiss-greenstone relationships in the Ancient Gneiss Complex of southwestern Swaziland, southern Africa, and implications for early crustal evolution. Precambrian Res 67:109–139

Kröner A, Compston W, Williams IS (1989) Growth of early Archaean crust in the Ancient Gneiss Complex of Swaziland as revealed by single zircon dating. Tectonophysics 161:271–298

Kröner A, Byerly GR, Lowe DR (1991a) Chronology of early Archean granite-greenstone evolution in the Barberton Mountain Land, South Africa, based on precise dating by single grain zircon evaporation. Earth Planet Sci Lett 103:41–54

Kröner A, Wendt JI, Tegtmeyer AR, Milisenda C, Compston W (1991b) Geochronology of the Ancient Gneiss Complex, Swaziland, and implications for crustal evolution. In: Ashwal LD (ed) Two cratons and an orogen – excursion guidebook and review articles for a field workshop through selected archaean terrains of Swaziland, South Africa, and Zimbabwe. IGCP project 280. Department of Geology, University of the Witwatersrand, Johannesburg, pp 8–31

Kröner A, Jaeckel P, Brandl G (2000) Single zircon ages for felsic to intermediate rocks from the Pietersburg and Giyani greenstone belts and bordering granitoid orthogneisses, northern Kaapvaal Craton, South Africa. J Afr Earth Sci 30(4):773–793

Kröner A, Hegner E, Byerly GR, Lowe DR (1992) Possible terrain identification in the early Archaean Barberton greenstone belt, South Africa, using single zircon geochronology. EOS Trans Am Geophys Union Fall Meet suppl 73(43):616

Kröner A, Hegner E, Wendt JI, Byerly GR (1996) The oldest part of the Barberton granitoid-greenstone terrain, South Africa: evidence for crust formation between 3.5 and 3.7 Ga. Precambrian Res 78:105–124

Kruckenberg SC, Chamberlain KR, Frost CD, Frost BR (2001) One billion years of Archean crustal evolution: Black Rock Mountain, northeastern Granite Mountains, Wyoming. Geol Soc Am Abstr Programs 33(6):A-401

Kumar A, Bhaskar Rao YJ, Sivaraman TV, Gopalan K (1996) Sm–Nd ages of Archaean metavolcanics of the Dharwar craton, South India. Precambrian Res 80:205–216

Kump LR (2009) The rise of atmospheric oxygen. Nature 451:277–278

Kushiro I (1972) Effect of water on the composition of magmas formed at high pressures. J Petrol 13:311–334

Kusky TM (1998) Tectonic setting and terrain accretion of the Archean Zimbabwe craton. Geology 26:163–166

Kusky TM, Kidd WSF (1992) Remnants of an Archaean oceanic plateau, Belingwe greenstone belt, Zimbabwe. Geology 20:43–46

Kyte FT (2002) Tracers of extraterrestrial components in sediments and inferences for Earth's accretion history. Geol Soc Am Spec Pap 356:21–38

Kyte FT, Zhou L, Lowe DR (1992) Noble metal abundances in an early Archaean impact deposit. Geochim Cosmochim Acta 56:1365–1372

Kyte FT, Shukolyukov A, Lugmair GW, Lowe DR, Byerly GR (2003) Early Archaean spherule beds: chromium isotopes confirm origin through multiple impacts of projectiles of carbonaceous chondrite type. Geology 31:283–286

Kyte FT, Shukolyukov A, Hildebrand AR, Lugmair GW, Hanova J (2004) Initial Cr-isotopic and iridium measurements of concentrates from Late-Eocene cpx spherule deposits. In: Lunar and planetary science XXXV #1824. Lunar and Planetary Institute, Houston

Lambert RSJ, Holland JG (1976) Amitsoq gneiss geochemistry: preliminary observations. In: Windley BF (ed) Early history of the Earth. Wiley, London, pp 191–202

Langford FF, Morin JA (1976) The development of the Superior Province of northwestern Ontario by merging island arcs. Am J Sci 276:1023–1034

Lanier WP, Lowe DR (1982) Sedimentology of the Middle Marker (3.4 Ga), Onverwacht Group, Transvaal, South Africa. Precambrian Res 18:237–260

Layer PW, Kröner A, York D (1992) Pre-3000 Ma thermal history of the Archean Kaap Valley pluton, South Africa. Geology 20:717–720

LeCheminant AN, Heaman LM (1989) Mackenzie igneous events, Canada: Middle Proterozoic hotspot magmatism associated with ocean opening. Earth Planet Sci Lett 96:38–48

Leclair AD, Boily M, Berclaz A, Labbé JY, Lacoste P, Simard M, Maurice C (2006) 1.2 billion years of Archean evolution in the northeastern Superior Province. Geol Assoc Can Abstr 31:85

Levine J, Becker TA, Muller RA, Renne PR (2005) 40Ar/39Ar dating of Apollo 12 impact spherules. Geophys Res Lett 32(15):L15201

Lindsay JF, Brasier MD, McLoughlin N, Green OR, Fogel M, McNamara KM, Steele A, Mertzman SA (2003) Abiotic Earth – establishing a baseline for earliest life, data from the Archean of Western Australia. Lunar Planet Sci XXXIV:1137

Lipple SL (1975) Definitions of new and revised stratigraphic units of the eastern Pilbara Region. West Aust Geol Surv Ann Rep 1974:58–63

Liu DY, Nutman AP, Compston W, Wu JS, Shen QH (1992) Remnants of >3800 Ma crust in the Chinese part of the Sino-Korean craton. Geology 20:339–342

Longdoz B, Francois LM (1997) The faint young sun climatic paradox: influence of the continental configuration and of the seasonal cycle on the climatic stability. Global Planet Change 14:97–112

Lowe DR (1980) Stromatolites 3,400-Myr old from the Archaean of Western Australia. Nature 284:441–443

Lowe DR (1999a) Petrology and sedimentology of cherts and related silicified sedimentary rocks in the Swaziland Supergroup. In: Lowe DR, Byerly GR (eds) Geologic evolution of the Barberton Greenstone Belt, South Africa. Geological Society of America special paper 329. Geological Society of America, Boulder, pp 83–114

Lowe DR (1999b) Shallow-water sedimentation of accretionary lapilli-bearing strata of the Msauli Chert: evidence of explosive hydromagmatic komatiitic volcanism. In: Lowe DR, Byerly GR (eds) Geologic evolution of the Barberton Greenstone Belt, South Africa. Geological Society of America special paper 329. Geological Society of America, Boulder, pp 213–232

Lowe DR (2013) Crustal fracturing and chert dike formation triggered by large meteorite impacts, ca. 3.260 Ga, Barberton greenstone belt, South Africa. Geol Soc Am Bull 125:894–912

Lowe DR, Byerly GR (1986a) Archaean flow-top alteration zones formed initially in a low-temperature sulphate-rich environment. Nature 324:245–248

Lowe DR, Byerly GR (1986b) Early Archaean silicate spherules of probable impact origin, South Africa and Western Australia. Geology 14:83–86

Lowe DR, Byerly GR (1999) Stratigraphy of the west-central part of the Barberton Greenstone Belt, South Africa. In: Lowe DR, Byerly GR (eds) Geologic evolution of the Barberton Greenstone Belt, South Africa. Geological Society of America special paper 329. Geological Society of America, Boulder, pp 1–36

Lowe DR, Byerly GR (2007) An overview of the geology of the Barberton greenstone belt and vicinity: implications for early crustal development. In: Van Kranendonk MJ, Smithies RH, Bennett VC (eds) Earth's oldest rocks, vol 15, Developments in Precambrian geology. Elsevier, Amsterdam, pp 481–526

Lowe DR, Byerly GR (2010) Did the LHB end not with a bang but with a whimper? 41st Lunar Planet Science Conference 2563pdf

Lowe DR, Fisher WG (1999) Sedimentology, mineralogy, and implications of silicified evaporites in the Kromberg Formation, Barberton Greenstone Belt, South Africa. In: Lowe DR, Byerly GR (eds) Geologic evolution of the Barberton Greenstone Belt, South Africa. Geological Society of America special paper 329. Geological Society of America, Boulder, pp 167–188

Lowe DR, Knauth LP (1977) Sedimentology of the Onverwacht Group (3.4 billion years), Transvaal, South Africa, and its bearing on the characteristics and evolution of the early Earth. J Geol 85:699–723

Lowe D, Knauth LP (1978) The oldest marine carbonate ooids reinterpreted as volcanic accretionary lapilli: Onverwacht Group, South Africa. J Sediment Petrol 48:709–722

Lowe DR, Nocita B (1999) Foreland basin sedimentation in the Mapepe Formation, southern-facies Fig Tree Group. In: Lowe DR, Byerly GR (eds) Geologic evolution of the Barberton Greenstone Belt, South Africa. Geological Society of America special paper 329. Geological Society of America, Boulder, pp 233–258

Lowe DR, Byerly GR, Asaro F, Kyte FJ (1989) Geological and geochemical record of 3400 million year old terrestrial meteorite impacts. Science 245:959–962

Lowe DR, Byerly GR, Kyte FT, Shukolyukov A, Asaro F, Krull A (2003) Characteristics, origin, and implications of Archaean impact-produced spherule beds, 3.47–3.22 Ga, in the Barberton Greenstone Belt, South Africa: keys to the role of large impacts on the evolution of the early Earth. Astrobiology 3:7–48

Macgregor AM (1932) The geology of the country around Que-Que, Gwelo District. Geol Surv S Rhodesia Bull 20:113

Macgregor AM (1951) Some milestones in the Precambrian of southern Rhodesia (anniversary address by president). Proc Geol Soc S Afr IIV:xxvii–lxxiv

Mahadevan TM (2004) Continental evolution through time: new insights from southern Indian shield (Abstract). In: International workshop on tectonics and evolution of the Precambrian Southern Granulite Terrain, India and Gondwana Correlations, Feb. 2004. National Geophysical Research Institute, Hyderabad, pp 37–38

Maibam B, Goswami JN, Srinivasan R (2011) Pb–Pb zircon ages of Archaean metasediments and gneisses from the Dharwar craton, southern India: implications for the antiquity of the eastern Dharwar craton. J Earth Syst Sci 120:643–661

Manikyamba C, Kerrich R, Khanna TC, Keshav Krishna A, Satyanarayanan M (2008) Geochemical systematics of komatiite-tholeiite and adakitic-arc basalt associations: the role of a mantle plume and convergent margin in formation of the Sandur superterrain, Dharwar craton, India. Lithos 106:155–172

Maphalala RM, Kröner A (1993) Pb-Pb single zircon ages for the younger Archaean granitoids of Swaziland, Southern Africa. In: Geological survey and mines, international colloquium of African geology, Mbabane, pp 201–206

Martin H (1999) Adakitic magmas: modern analogues of Archaean granitoids. Lithos 46:411–429

Martin H, Moyen JF (2002) Secular changes in tonalite-trondhjemite-granodiorite composition as markers of the progressive cooling of the Earth. Geology 30:319–322

Martin H, Smithies RH, Rapp R, Moyen JF, Champion D (2005) An overview of adakite, tonalite–trondhjemite–granodiorite (TTG), and sanukitoid: relationships and some implications for crustal evolution. Lithos 79:1–24

Martin W, Baross J, Kelley D, Russell MJ (2008) Hydrothermal vents and the origin of life. Nat Rev Microbiol 6:805–814

Mathur SP (1976) Relation of Bouguer anomalies to crustal structure in southwestern and central Australia. Bur Miner Resour J Aust Geol Geophys 1:277–286

Matthews PE, Charlesworth EG, Eglington BM, Harmer RE (1989) A minimum 3.29 Ga age for the Nondweni greenstone complex in the southeastern Kaapvaal Craton. South Afr J Geol 92:272–278

McCarthy T, Robb J (1978) On the relationship between cumulus mineralogy and trace and alkali element chemistry in an Archaean granite from the Barberton region, South Africa. Geochim Cosmochim Acta 42:21–26

McCollom TM, Seewald JS (2006) Carbon isotope composition of organic compounds produced by abiotic synthesis under hydrothermal conditions. Earth Planet Sci Lett 243:74–84

McCulloch MT, Bennett VC (1994) Progressive growth of the Earth's continental crust and depleted mantle: geochemical constraints. Geochim Cosmochim Acta 58:4717–4738

McCulloch MT, Gamble JA (1991) Geochemical and geodynamical constraints on subduction zone magmatism. Earth Planet Sci Lett 102:358–374

McDonough WF, Ireland TR (1993) Intraplate origin of komatiites inferred from trace elements in glass inclusions. Nature 365:432–434

McDonough WE, Sun S (1995) The composition of the Earth. Chem Geol 120:223–253

McGregor VR (1973) The early Precambrian geology of the Godthab District, west Greenland. Philos Trans R Soc Lond A 273:343–358

McGregor VR, Mason B (1977) Petrogenesis and geochemistry of metabasaltic and metasedimentary enclaves in the Amitsoq gneisses, west Greenland. Am Mineral 62:887–904

McNaughton NJ, Green MD, Compston W, Williams IS (1988) Are anorthositic rocks basement to the Pilbara Craton? Geol Soc Aust Abstr 21:272–273

Meen JK, Rogers JJW, Fullagar PD (1992) Lead isotopic compositions of the western Dharwar craton, southern India: evidence for distinct middle Archaean terrains in late Archaean craton. Geochim Cosmochim Acta 56:2455–2470

Melosh HJ, Vickery AM (1991) Melt droplet formation in energetic impact events. Nature 350:494–497

Mitroff II (1974) The subjective side of science. Elsevier, Amsterdam, 329 pp

Mogk DW, Mueller PA, Wooden J (1992) The significance of Archean terrain boundaries: evidence from the northern Wyoming province. Precambrian Res 55:155–168

Mogk DW, Mueller PA, Wooden JL (2004) Tectonic implications of late Archeanearly Proterozoic supracrustal rocks in the Gravelly Range, SW Montana. In: Geological Society of America 2004 Annual Meeting Abstract with Programs 36(5). p 507

Mohan MR et al (2014) SHRIMP zircon and titanite U-Pb ages, Lu-Hf isotope signatures and geochemical constraints for ~2.56 Ga granitic magmatism in western Dharwar Craton, southern India: evidence for short-lived Neoarchean episodic crustal growth. Precambrian Res Online 6 January 2014

Mojzsis SJ (2007) Sulphur on the early Earth. In: Van Kranendonk MJ, Smithies RH, Bennett VC (eds) Earth's oldest rocks, vol 15, Developments in Precambrian geology. Elsevier, Amsterdam, pp 923–970

Mojzsis SJ, Harrison TM (2000) Vestiges of a beginnings: clues to the emergent biosphere recorded in the oldest known rocks. Geol Soc Am Today 10:1–6

Mojzsis SJ, Harrison TM, Pidgeon RT (2001) Oxygen-isotope evidence from ancient zircons for liquid water at the Earth's surface 4,300 Myr ago. Nature 409:178–180

Mozjsis SJ, Arrhenius G, McKeegan KD, Harrison TM, Friend CRL (1996) Evidence for life on Earth before 3800 million years ago. Nature 270: 43–45

Moorbath S (1977) Ages, isotopes and the evolution of the Precambrian continental crust. Chem Geol 20:151–187

Moorbath S, O'Nions RK, Pankhurst RJ (1973) Early Archaean age for the Isua Iron Formation, West Greenland. Nature 245:138–139

Moorbath S, Wilson JF, Cotterill P (1976) Early Archaean age for the Sebakwian Group at Selukwe, Rhodesia. Nature 264:536–538

Morey GB, Sims PK (1976) Boundary between two lower Precambrian terrains in Minnesota and its geological significance. Bull Geol Soc Am 87:141–152

Morris RC (1993) Genetic modeling for banded iron-formation of the Hamersley Group, Pilbara Craton, Western Australia. Precambrian Res 60:243–286

Mory AJ, Iasky RP, Glikson AY, Pirajno F (2000) Woodleigh Carnarvon basin, Western Australia: a new 120 km-diameter impact structure. Earth Planet Sci Lett 177:119–128

Moyen JF, Martin H (2012) Forty years of TTG research. Lithos 148:312–336

Moyen JF, Stevens G, Kisters AFM, Belcher RW (2007) TTG plutons of the Barberton granitoid-greenstone terrain, South Africa. In: Van Kranendonk MJ, Smithies RH, Bennett VC (eds) Earth's oldest rocks. Developments in Precambrian geology 15. Elsevier, Amsterdam, pp 607–667

Moyen JF, Champion D, Smithies RH (2009) The geochemistry of Archaean plagioclase-rich granites as a marker of source enrichment and depth of melting. Trans R Soc Edinb 100:35–50

Moyen JF, Smithies RH, Champion DC (2011) Granitoids in the East Pilbara craton and the Barberton Granite-Greenstone terrain record different modes of Archaean crustal accretion. Seventh Hutton symposium on granites and related rocks, Avila, Spain, July 2011

Mueller PA, Frost CD (2006) The Wyoming province: a distinctive Archean craton in Laurentian North America. Can J Earth Sci 43:1391–1397

Mueller PA, Heatherington AL, Kelly DM, Wooden JL, Mogk DW (2002) Paleoproterozoic crust within the Great Falls tectonic zone; implications for the assembly of southern Laurentia. Geology 30:127–130

Myers JS (1988) Early Archaean Narryer Gneiss Complex, Yilgarn Craton, Western Australia. Precambrian Res 38:297–307

Myers JS (1990) Western Gneiss Terrain. In: Geology and mineral resources of Western Australia. Western Australia Geological Survey Memoir 3. pp 13–31

Myers JS (1997) Archaean geology of the Eastern Goldfields of Western Australia; regional overview. Precambrian Res 83:1–10

Myers JS, Swager C (1997) The Yilgarn Craton. In: de Wit M, Ashwal LD (eds) Greenstone belts. Clarendon, Oxford, pp 640–656

Mysen BO, Boettcher AL (1975) Melting of a hydrous mantle, I – phase relations of natural periodotite at high pressures and temperatures with controlled activities of water, CO2 and H. J Petrol 16:520–548

Naqvi SM (1976) Physical – chemical conditions during the Archaean as indicated by Dharwar geochemistry. In: Windley BF (ed) Early history of the Earth. Wiley, London, pp 289–298

Naqvi SM (1981) The oldest supracrustals of the Dharwar Craton, India. J Geol Soc India 22:458–469

Naqvi SM, Rogers JJW (1987) Precambrian geology of India. Oxford University Press, New York, 223 p

Naqvi SM, Divakara Rao, Narain H (1978) The primitive crust: evidence from the Indian Shield. Precambrian Res 6:323–346

Nelson DR (1997) Evolution of the Archaean granite-greenstone terrains of the Eastern Goldfields, Western Australia. SHRIMP U-Pb zircon constraints. Precambrian Res 83:57–81

Nelson DR (1999) Compilation of SHRIMP U-Pb Zircon Geochronology Data, 1998. Western Australia Geological Survey Record 1999/2

Nelson DR (2008) Geochronology of the Archean of Australia. Aust J Earth Sci 55:779–793, 55

Nemchin AA, Pidgeon RT, Whitehouse MJ (2006) Re-evaluation of the origin and evolution of >4.2 Ga zircons from the Jack Hills metasedimentary rocks. Earth Planet Sci Lett 244:218–233

Noffke N, Christian D, Wacey D, Hazen RM (2013) Microbially induced sedimentary structures recording an ancient ecosystem in the ca. 3.48 billion-year-old Dresser Formation, Pilbara, Western Australia. Astrobiology 13(12):1103–1124

Noldart AJ, Wyatt JD (1962) The geology of a portion of the Pilbara Goldfield, covering the Marble Bar and Nullagine 4-mile map sheets. Western Australia Geological Survey Bulletin 115:199 pp

Nutman AP, Friend CRL (2006) Re-evaluation of oldest life evidence: infrared absorbance spectroscopy and petrography of apatites in ancient metasediments, Akilia, W. Greenland. Precambrian Res 147:100–106

Nutman AP, Bennett V, Kinny PD, Price R (1993) Large-scale crustal structure of the northwestern Yilgarn Craton, Western Australia: evidence from Nd isotopic data and zircon geochronology. Tectonics 12:971–981

Nutman AP, Chadwick B, Krishna Rao B, Vasudev VN (1996) SHRIMP U–Pb zircon ages of acid volcanic rocks in the Chitradurga and Sandur Groups and granites adjacent to Sandur schist belt. J Geol Soc India 47:153–161

Nutman AP, Bennett VC, Friend CRL, Rosing MT (1997) ~3710 and 3790 Ma volcanic sequences in the Isua (Greenland) supracrustal belt; structural and Nd isotope implications. Chem Geol 141:271–287

Nutman AP, Friend CRL, Horie K, Hidaka H (2007) The Itsaq gneiss complex of southern west Greenland and the construction of Archaean crust at convergent plate boundaries. In: Van Kranendonk MJ, Smithies RH, Bennett VC (eds) Earth's oldest rocks, vol 15, Developments in Precambrian geology. Elsevier, Amsterdam, pp 187–218

Nutman AP, Clark Friend RL, Bennett VC, Wright D, Norman MD (2010) ≥3700 Ma premetamorphic dolomite formed by microbial mediation in the Isua supracrustal belt (W. Greenland): simple evidence for early life? Precambrian Res 183:725–737

O'Beirne WR (1968) The acid porphyries and porphyroid rocks of the Kalgoorlie area. Ph.D. thesis, University of Western Australia

Oberthür T, Davis DW, Blenkinsop TG, Höhndorf A (2002) Precise U–Pb mineral ages, Rb–Sr and Sm–Nd systematics for the Great Dyke, Zimbabwe – constraints on late Archean events in the Zimbabwe Craton and Limpopo Belt. Precambrian Res 113:293–305

O'Connor JT (1965) A classification of quartz rich igneous rocks based on feldspar ratios. US Geological Survey Professional paper 525B

O'Hara MJ (1977) Thermal history of excavation of Archaean gneisses from the base of the continental crust. J Geol Soc Lond 134:185–200

Ohmoto H, Watanabe Y, Ikemi H, Poulson SR, Taylor BE (2006) Sulphur isotope evidence for an oxic Archaean atmosphere. Nature 442:908–911

Ohta H, Maruyama S, Takahashi E, Watanabe Y, Kato Y (1996) Field occurrence, geochemistry and petrogenesis of the Archaean Mid-Oceanic Ridge Basalts (A-MORBs) of the Cleaverville area, Pilbara Craton, Western Australia. Lithos 37:199–221

O'Keefe JD, Aherns TJ (1982) Interaction of the Cretaceous/Tertiary extinction bolide with the atmosphere. Geological Society of America, special paper 190. pp 103–120

Olivarez AM, Owen RM (1991) The europium anomaly of seawater: implications for fluvial versus hydrothermal REE inputs to the oceans. Chem Geol 92:317–328

O'neill J, Maurice C, Stevenson RK, Larocque J, Cloquet C, David J, Francis D (2007) The geology of the 3.8 Ga Nuvvuagittuq (Porpoise Cove) Greenstone Belt, Northeastern Superior Province, Canada. In: Van Kranendonk MJ, Smithies RH, Bennett VC (eds) Earth's oldest rocks, vol 15, Developments in Precambrian geology. Elsevier, Amsterdam, pp 219–250

O'Neill JM, Lopez DA (1985) Character and regional significance of Great Falls Tectonic Zone, east-central Idaho and west-central Montana. Am Assoc Petrol Geol Bull 69:437–447

O'Nions RK, Pankhurst RJ (1974) Rare earth elements distribution in Archaean gneisses and anorthosites, Godthab area, west Greenland. Earth Planet Sci Lett 22:328–338

O'Nions RK, Pankhurst RJ (1978) Early Archaean rocks and geochemical evolution of the Earth's crust. Earth Planet Sci Lett 38:211–236

O'Reilly SY, Griffin WL, Pearson NJ, Jackson SE, Belousova EA, Alard O, Saeed A (2008) Taking the pulse of the Earth: linking crustal and mantle events. Aust J Earth Sci 55:983–996

Oversby VM (1976) Isotopic ages and geochemistry of Archaean acid igneous rocks from the Pitbara, Western Australia. Geochim Cosmochim Acta 40:817–829

Papanastassiodu A, Wasserburg GJ (1969) Initial Sr isotopic abundances and the resolution of small time differences in the formation of planetary objects. Earth Planet Sci Lett 5:361–376

Pavlov AA, Kasting JF (2002) Mass-independent fractionation of sulfur isotopes in Archean sediments: strong evidence for an anoxic Archean atmosphere. Astrobiology 2:27–41

Pearce JA, Cann JR (1971) Ophiolite origin investigated through discriminant analysis using Ti, Zr and Y. Earth Planet Sci Lett 12:339–349

Pearce JA, Cann JR (1973) Tectonic setting of basic volcanic rocks determined using trace element analysis. Earth Planet Sci Lett 19:290–300

Pearce JA, Gorman BE, Birkett TC (1977) The relationships between major element geochemistry and tectonic environments of basic and intermediate volcanic rocks. Earth Planet Sci Lett 36:121–132

Peck WH, Valley JW, Wilde SA, Graham CM (2001) Oxygen isotope ratios and rare earth elements in 3.3 to 4.4 Ga zircons: ion microprobe evidence for high $\delta^{18}O$ continental crust and oceans in the Early Archaean. Geochim Cosmochim Acta 65:4215–4229

Percival JA (2007) Eo to MesoArchaean terrains of the superior province and their tectonic context. In: Van Kranendonk MJ, Smithies RH, Bennett VC (eds) Earth's oldest rocks. Developments in Precambrian geology, vol 15, Elsevier, Amsterdam, pp 1065–1085

Percival J, Card KD (1994) Geology, Lac Minto – Rivière aux Feuilles. Geological Survey of Canada Map 1854A (1:500,000)

Percival JA, Mortensen JK, Stern RA, Card KD, Bégin NJ (1992) Giant granulite terrains of northeastern Superior Province: the Ashuanipi complex and Minto block. Can J Earth Sci 29:2287–2308

Percival JA, McNicoll V, Brown JL, Whalen JB (2004) Convergent margin tectonics, central Wabigoon subprovince, Superior Province, Canada. Precambrian Res 132:213–244

Percival JA, Sanborn-Barrie M, Skulski T, Stott GM, Helmstaedt H, White DJ (2006) Tectonic evolution of the western Superior Province from NATMAP and Lithoprobe studies. Can J Earth Sci 43:1085–1117

Perry EC, Ahmed SN (1977) Carbon isotope composition of graphite and carbonate minerals from 3.8-AE metamorphosed sediments, Isukasia, Greenland. Earth Planet Sci Lett 36:280–284

Phaup AE (1973) The granitic rocks of the Rhodesian craton. Geol Soc S Afr Spec Publ 3:59–68

Pichamuthu CS (1960) Charnockite in the making. Nature 188:135–136

Pichamuthu CS (1968) The Precambrian of India. In: Rankama K (ed) The Precambrian, vol 3. Interscience, New York, pp 1–96

Pidgeon RT (1978) Geochronological investigations of granite batholiths of the Archaean granite – greenstone terrain of the Pilbara Block. Western Australia. In: Smith IEM, Williams JG (eds) Proceedings of the 1978 Archaean geochemistry conference. University of Toronto, Ontario, pp 360–362

Pidgeon RT, Hallberg JA (2000) Age relationships in supracrustal sequences in the northern part of the Murchison Terrain, Archaean Yilgarn Craton, Western Australia: a combined field and zircon U-Pb study. Aust J Earth Sci 47:153–165

Pidgeon RT, Nemchin AA (2006) High abundance of early Archaean grains and the age distribution of detrital zircons in a sillimanite-bearing quartzite from Mt Narryer, Western Australia. Precambrian Res 150:201–220

Pidgeon RT, Wilde SA (1990) The distribution of 3.0 Ga and 2.7 Ga volcanic episodes in the Yilgarn Craton of Western Australia. Precambrian Res 48:309–325

Pidgeon RT, Wingate TD, Bodorkos S, Nelson DR (2010) The age distribution of detrital zircons in quartzites from the Toodyay-Lake Grace domains, Western Australia: implications for the early evolution of the Yilgarn. Am J Sci 310:1115–1135

Pike G, Cas RAF (2002) Stratigraphic evolution of Archaean volcanic rock-dominated rift basins from the Whim Creek Belt, west Pilbara Craton, Western Australia. In: Altermann W, Corcoran P (eds) Precambrian sedimentary environments: a modern approach to depositional systems. International Association of Sedimentologists Special publication 33. Blackwell Science, Oxford, pp 213–234

Pirajno F (2000) Ore deposits and mantle plumes. Kluwer Academic Publishers, Dordrecht, 556 pp

Pirajno F (2007a) Mantle plumes, associated intraplate tectono-magmatic processes and ore systems. Episodes 30:6–19

Pirajno F (2007b) Ancient to modern Earth: the role of mantle plumes in the making of continental crust. In: Van Kranendonk MJ, Smithies RH, Bennett VC (eds) Earth's oldest rocks. Developments in Precambrian geology 15. Elsevier, Amsterdam, pp 1037–1064

Polat A (2013) Geochemical variations in Archean volcanic rocks, southwestern Greenland: traces of diverse tectonic settings in the early Earth. Geology 41:379–380

Polat A, Hofmann AW, Rosing MT (2002) Boninite-like volcanic rocks in the 3.7–3.8 Ga Isua greenstone belt, West Greenland: geochemical evidence for intra-oceanic subduction zone processes in the early Earth. Chem Geol 184:231–254

Pope KO, Baines KH, Ocampo AC, Ivanov BA (1997) Energy volatile production and climatic effects of the Chicxulub Cretaceous/Tertiary impact. J Geophys Res 102:21645–21664

Poujol M (2007) An overview of the pre-mesoarchaean rocks of the Kaapvaal Craton, South Africa. In: Van Kranendonk MJ, Smithies RH, Bennett VC (eds) Earth's oldest rocks, vol 15, Developments in Precambrian geology. Elsevier, Amsterdam, pp 453–463

Poujol M, Anhaeusser CR (2001) The Johannesburg Dome, South Africa: new single zircon U-Pb isotopic evidence for early Archaean granite-greenstone development within the Central Kaapvaal Craton. Precambrian Res 108:139–157

Poujol M, Robb LJ, Respaut JP, Anhaeusser CR (1996) 3.07–2.97 Ga greenstone belt formation in the northeastern Kaapvaal Craton: implications for the origin of the Witwatersrand Basin. Econ Geol 91:1455–1461

Poujol M, Robb LJ, Anhaeusser CR, Gericke B (2003) A review of the geochronological constraints on the evolution of the Kaapvaal Craton, South Africa. Precambrian Res 127:181–213

Pross A (2004) Causation and the origin of life: metabolism or replication first? Orig Life Evol Bios 34(3):307–321
Radhakrishna BP (1975) The two greenstone groups in the Dharwar Craton. Indian Miner 16:12–15
Ramakrishnan M, Vaidyanadhan R (2008) Geology of India, vol 1. Geol. Soc. India, Bangalore, 556 p
Ramiengar AS, Ramakrishnan M, Vishwanatha MN (1978) Charnockite-gneiss complex relationship in southern Karanataka. J Geol Soc India 19:411–419
Ransom B, Byerly GR, Lowe DR (1999) Subaqueous to subaerial Archean ultramafic phreatomagmatic volcanism, Kromberg Formation, Barberton Greenstone Belt, South Africa. In: Lowe DR, Byerly GR (eds) Geologic evolution of the Barberton Greenstone Belt, South Africa. Geological Society of America special paper 329. Geological Society of America, Boulder, pp 151–166
Rasmussen B, Koeberl C (2004) Iridium anomalies and shocked quartz in a Late Archean spherule layer from the Pilbara Craton: new evidence for a major asteroid impact at 2.63 Ga. Geology 32:1029–1032
Rasmussen B, Blake TS, Fletcher IR (2005) U-Pb zircon age constraints on the Hamersley spherule beds: evidence for a single 2.63 Ga Jeerinah-Carawine impact ejecta layer. Geology 33:725–728
Ravindra Kumar GR, Raghavan V (1992) The incipient charnockites in transition zone, khondalite zone and granulite zones of South India: controlling factors and contrasting mechanisms. J Geol Soc India 39:293–302
Reimold WU, Meyer FM, Walraven F, Matthews PE (1993) Geochemistry and chronology of the pre- and post-Pongola granitoids for northeastern Transvaal. In: Geological survey and mines, 16th colloquium of African geology, Mbabane, Swaziland, pp 294–296
Reynolds DG, Brook WA, Marshall AE, Allchaurch PD (1975) Volcanogenic copper-zinc deposits in the Pilbara and Yilgarn Archaean Blocks. In: Knight CL (ed) Economic geology of Australia and Papua New Guinea. Australasian Institute of Mining and Metallurgy Monograph 5. Australasian Institute of Mining and Metallurgy, Parkville, pp 185–195
Ringwood AE (1975) Composition and petrology of the Earth's mantle. McGraw-Hill, New York
Ringwood AE (1986) Origin of the Earth and Moon. Nature 322:323–328
Robb LJ, Anhauesser CR (1983) Chemical and petrogenetic characteristics of Archean tonalite-trondhjemite gneiss plutons in the Barberton Mountain Land. In: Anhauesser CR (ed) Contributions to the geology of the Barberton Mountain Land. Geological Society of South Africa special publication 9. Geological Society of South Africa, Johannesburg, pp 103–116
Roberts JA, Bennett PC, González LA, Macpherson GL, Milliken KL (2004) Microbial precipitation of dolomite in methanogenic groundwater. Geology 32:277–280
Rose NM, Rosing M, Bridgwater D (1996) The origin of metacarbonate rocks in the Archean Isua supracrustal belt, West Greenland. Am J Sci 296:1004–1044
Rosen OM, Turkina OM (2007) The oldest rock assemblage of the Siberian Craton. In: Van Kranendonk MJ, Smithies RH, Bennett VC (eds) Earth's oldest rocks, vol 15, Developments in Precambrian geology. Elsevier, Amsterdam, pp 793–838
Rosing MT (1999) 13C-depleted carbon microparticles in >3700-Ma sea-floor sedimentary rocks from West Greenland. Science 283:674–676
Rosing MT, Bird DK, Sleep NH, Bjerrum CJ (2010) No climate paradox under the faint early Sun. Nature 464:744–749
Royer DL, Berner RA, Beerling DJ (2001) Phanerozoic atmospheric CO change: evaluating geochemical and paleobiological approaches. Earth Sci Rev 54:349–392
Royer DL, Berner RA, Park J (2007) Climate sensitivity constrained by CO2 concentrations over the past 420 million years. Nature 446:530–532
Royer DL, Berner RA, Montañez I, Neil P, Tabor J, Beerling DJ (2004) CO2 as a primary driver of Phanerozoic climate. Geol Soc Am Today 14:3
Ruddiman WF (1997) Tectonic uplift and climate change. Plenum Press, New York, 535 pp
Ruddiman WF (2003) Orbital insolation, ice volume, and greenhouse gases. Quat Sci Rev 22:1597–1629

Russell M (ed), Lane N, Trifonov EN, Rampelotto PH, Freeland S (2010) Origins, Abiogenesis and the search for life in the universe. Cosmology Science Publishers. 500 pp. http://www.amazon.com/Origins-Abiogenesis-Search-Life-Universe/dp/0982955219

Russell MJ, Hall AJ (2006) The onset and early evolution of life. Geol Soc Am Mem 198:32

Ryder G (1990) Lunar samples, lunar accretion and the early bombardment of the Moon. EOS Trans Am Geophys Union 71:313–322

Ryder G (1991) Accretion and bombardment in the Earth–Moon system: the lunar record. Lunar Planet Sci Inst Contrib 746:42–43

Ryder G (1997) Coincidence in the time of the Imbrium Basin impact and Apollo 15 Kreep volcanic series: impact induced melting? Lunar Planet Sci Inst Contrib 790:61–62

Sagan C, Mullen G (1972) Earth and mars: evolution of atmospheres and surface temperatures. Science 177:52–56

Sakurai R, Ito M, Ueno Y, Kitajima K, Maruyama S (2005) Facies architecture and sequence-stratigraphic features of the Tumbiana Formation in the Pilbara Craton, northwestern Australia: implications for depositional environments of oxygenic stromatolites during the late Archean. Precambrian Res 138:255–273

Sano Y, Terada K, Hidaka H, Yokoyama K, Nutman AP (1999) Palaeoproterozoic thermal events recorded in the ~4.0 Ga Acasta gneiss, Canada: evidence from SHRIMP U-Pb dating of apatite and zircon. Geochim Cosmochim Acta 63:899–905

Scherer E, Munker C, Mezger K (2001) Calibration of the lutetium-hafnium clock. Science 293:683–687

Schidlowski M, Appel PWU, Eichmann R, Junge CE (1979) Carbon isotope geochemistry of the 3.7×10^9-yr old Isua sediments, West Greenland: implications for the Archaean carbon and oxygen cycles. Geochim Cosmochim Acta 43:189–199

Schmitz MD, Bowring SA, de Wit M, Gartz V (2004) Subduction and terrain collision stabilize the western Kaapvaal craton tectosphere 2.9 billion years ago. Earth Planet Sci Lett 222:363–376

Schoenberg R, Kamber B, Collerson KD, Moorbath (2002) Tungsten isotope evidence from ~3.8 Gyr metamorphosed sediments for early meteorite bombardment of the Earth. Nature 418:403–405

Schoene B, Dudas FOL, Bowring SA, DeWit M (2009) Sm–Nd isotopic mapping of lithospheric growth and stabilization in the eastern Kaapvaal craton. Terra Nova 21:219–228

Schopf JW (2001) Cradle of life: the discovery of Earth's earliest fossils. Princeton University Press, Princeton

Schopf JW, Packer BM (1987) Early Archaean (3.3-billion to 3.5-billion-year-old) microfossils from Warrawoona Group, Australia. Science 237:70–73

Schopf JW, Kudryavtsev AB, Czaja AD, Tripathi AB (2007) Evidence of Archean life: stromatolites and microfossils. Precambrian Res 158:141–155

Shackleton RM (1976) Shallow and deep level exposures of the Archaean crust in India and Africa. In: Windley BF (ed) Early history of the Earth. Wiley, London, pp 317–322

Sharma RS (2009) Cratons and fold belts of India. Lecture notes in earth sciences 127. Springer, Berlin, pp 41–115

Sharma M, Basu AR, Ray SL (1994) Sm-Nd isotopic and geochemical study of Archaean tonalitic-amphibolite association from the eastern Indian craton. Contrib Mineral Petrol 117:45–55

Shen QH, Geng YS, Song B, Wan YS (2005) New information from surface outcrops and deep crust of Archaean Rocks of the North China and Yangtze Blocks, and Qinling-Dabie Orogenic Belt. Acta Geologica Sinica 79:616–627

Sheraton JW (1970) The origin of the Lewisian gneisses of northwest Scotland, with particular reference to the Drumbeg area, Sutherland. Earth Planet Sci Lett 8:301–310

Shirey SB, Walker RJ (1998) The Re-Os system in cosmochemistry and high-temperature geochemistry. Ann Rev Earth Planet Sci 26:423–500

Shirey SB, Wilson AH, Carlson RW (1998) Re-Os isotopic systematics of the 3300 Ma Nondweni greenstone belt, South Africa: implications for Komatiite Formation at a Craton Edge. In: AGU Spring Meeting, Boston

Shoemaker EM, Shoemaker CS (1996) The Proterozoic impact record of Australia. Aust Geol Surv Org J Aust Geol Geophys 16:379–398

Shukolyukov A, Kyte FT, Lugmair GW, Lowe DR, Byerly GR (2000) The oldest impact deposits on Earth. In: Koeberl C, Gilmour I (eds) Lecture notes in Earth science 92: impacts and the early Earth. Springer, Berlin, pp 99–116

Simonson BM (1992) Geological evidence for an early Precambrian microtektite strewn field in the Hamersley Basin of Western Australia. Geol Soc Am Bull 104:829–839

Simonson BM, Carney KE (1999) Roll-Up Structures: evidence of in situ microbial mats in Late Archean deep shelf environments. Palaios 14:13–24

Simonson BM, Glass BP (2004) Spherule layers – records of ancient impacts. Ann Rev Earth Planet Sci 32:329–361

Simonson BM, Hassler SW (1997) Revised correlations in the early Precambrian Hamersley Basin based on a horizon of re-sedimented impact spherules. Aust J Earth Sci 44:37–48

Simonson BM, Schubel KA, Hassler SW (1993) Carbonate sedimentology of the early Precambrian Hamersley Group of Western Australia. Precambrian Res 60:287–335

Simonson BM, Davies D, Wallace M, Reeves S, Hassler SW (1998) Iridium anomaly but no shocked quartz from Late Archaean microkrystite layer: oceanic impact ejecta? Geology 26:195–198

Simonson BM, Hassler SW, Beukes N (1999) Late Archaean impact spherule layer in South Africa that may correlate with a layer in Western Australia. In: Dressler BO, Sharpton VL (eds) Impact cratering planetary evolution. Geological Society of America special paper 339. Geological Society of America, Boulder, pp 249–262

Simonson BM, Davies D, Hassler SW (2000a) Discovery of a layer of probable impact melt spherules in the late Archaean Jeerinah Formation, Fortescue Group, Western Australia. Aust J Earth Sci 47:315–325

Simonson BM, Koeberl C, McDonald I, Reimold WU (2000b) Geochemical evidence for an impact origin for a late Archaean spherule layer Transvaal supergroup South Africa. Geology 28:1103–1106

Simonson BM, Cardiff M, Schubel KA (2001) New evidence that a spherule layer in the late Archaean Jeerinah Formation of Western Australia was produced by a major impact. In: 32nd lunar planetary science conference abstracts, lunar and planetary institute contribution 1080, Huston

Simonson BM, McDonald I, Shukolyukov A, Koeberl C, Reimold WU, Lugmair GW (2009a) Geochemistry of 2.63–2.49 Ga impact spherule layers and implications for stratigraphic correlations and impact processes. Precambrian Res 175:51–76

Simonson BM, Sumner DY, Beukes NJ, Johnson S, Gutzmerd J (2009b) Correlating multiple Neoarchean–Paleoproterozoic impact spherule layers between South Africa and Western Australia. Precambrian Res 169:100–111

Singh AP, Mishra DC, Vijaya Kumar Vyaghreswara Rao MBS (2003) Gravity magnetic signatures and crustal architecture along Kuppam-Palani geotransect, south India. Geol Soc India Mem 50:139–163

Skulski T, Corkery MT, Stone D, Whalen JB, Stern RA (2000) Geological and geochronological investigations in the Stull Lake – Edmund Lake greenstone belt and granitoid rocks of the northwestern Superior Province. In: Report of activities 2000, Manitoba Industry, Trade and Mines, Manitoba Geological Survey. pp 117–128

Slabunov AI et al (2006) The Archaean of the Baltic Shield: geology, geochronology, and geodynamic settings. Geotectonics 40:409–433

Smit J, Klaver G (1982) Sanidine spherules of the K-T boundary indicate a large impact event. Nature 292:47–49

Smithies RH (2000) The Archaean tonalite–trondhjemite–granodiorite (TTG) series is not an analogue of Cenozoic adakite. Earth Planet Sci Lett 182:115–125

Smithies RH (2002) Archaean boninite-like rocks in an intra-cratonic setting. Earth Planet Sci Lett 197:19–34

Smithies RH (2004) Geology of the De Grey and Pardoo 1:100 000 Sheets. Geological Survey of Western Australia 1:100 000 Geological series explanatory notes, Perth

Smithies RH, Champion DC (1999) Late Archaean felsic alkaline igneous rocks in the Eastern Goldfields, Yilgarn Craton, Western Australia: a result of lower crustal delamination? J Geol Soc 156:561–576

Smithies RH, Champion DC (2000) The Archaean high-Mg diorite suite: links to tonalite-trondhjemite-granodiorite magmatism and implications for early Archaean crustal growth. J Petrol 41:1653–1671

Smithies RH, Champion DC, Cassidy KF (2003) Formation of Earth's early Archaean continental crust. Precambrian Res 127:89–101

Smithies RH, Van Kranendonk MJ, Champion DC (2005) It started with a plume. Earth Planet Sci Lett 238:284–297

Smithies RH, Champion DC, Van Kranendonk MJ, Hickman AH (2007a) Geochemistry of volcanic rocks of the northern Pilbara Craton, Western Australia. Geological Survey of Western Australia Report 104

Smithies RH, Champion DC, Van Kranendonk MJ (2007b) The oldest well-preserved felsic volcanic rocks on Earth: geochemical clues to the early evolution of the Pilbara supergroup and implications for the growth of a paleoarchaean protocontinent. In: Van Kranendonk MJ, Smithies RH, Bennett VC (eds) Earth's oldest rocks. Developments in Precambrian geology, vol 15. pp 339–366

Smithies RH, Champion DC, Van Kranendonk MJ (2009) Formation of Paleoarchean continental crust through the infracrustal melting of enriched basalt. Earth Planet Sci Lett 281:298–306

Song B, Nutman AP, Liu D, Wu J (1996) 3800 to 2500 Ma crustal evolution in the Anshan area of Lianoning Province, NW China. Precambrian Res 78:79–94

Srinivasan R, Sreenivas BL (1972) Dharwar stratigraphy. J Geol Soc India 13:75–85

Stauffer MR, Mukherjee C, Koo J (1975) The Amisk Group: an Aphebian (?) island arc deposit. Can J Earth Sci 12:2021–2035

Stevenson DJ (1987) Origin of the Moon – the Collison Hypothesis 1987. Ann Rev Earth Planet Sci 15:271–315

Stevenson RK, Patchett PJ (1990) Implications for the evolution of continental crust from Hf isotope systematics of Archean detrital zircons. Geochim Cosmochim Acta 54:1683–1697

Stevenson RK, David J, Parent M (2006) Crustal evolution in the western Minto Block, northern Superior Province, Canada. Precambrian Res 145:229–242

Stott GM (1997) The Superior Province, Canada. In: de Wit MJ, Ashwal LD (eds) Greenstone belts. Oxford monograph on geology and geophysics, 35. Clarendon, Oxford, pp 480–507

Stott GM, Corfu F (1991) Uchi subprovince. In: Thurston PC et al (eds) Geology of Ontario. Ontario Geological Survey special volume 4. pp 145–238

Stowe CW (1973) The older tonalite gneiss complex in the Selukwe area, Rhodesia. Geol Soc S Afr Spec Publ 3:85–96

Strik G, de Wit MJ, Langereis CG (2007) Palaeomagnetism of the Neoarchaean Pongola and Ventersdorp supergroups and an appraisal of the 3.0–1.9 Ga apparent polar wander path of the Kaapvaal Craton, Southern Africa. Precambrian Res 153:96–115

Sugitania K, Grey K, Nagaokac T, Mimurad K, Walter M (2009) Taxonomy and biogenicity of Archaean spheroidal microfossils (ca 3.0 Ga) from the Mount Goldsworthy–Mount Grant area in the northeastern Pilbara Craton, Western Australia. Precambrian Res 173:50–59

Sumner DY, Beukes NJ (2006) Sequence stratigraphic development of the Neoarchean Transvaal carbonate platform, Kaapvaal Craton, South Africa. S Afr J Geol 109:11–22

Sun SS (1987) Chemical composition of Archaean komatiites: implications for early history of the earth and mantle evolution. J Volcanol Geotherm Res 32:67–82

Sun SS, Hickman AH (1998) New Nd-isotopic and geochemical data from the west Pilbara: implications for Archaean crustal accretion and shear zone development. Aust Geol Surv Org Res New 28:25–29

Sun SS, McDonough WF (1989) Chemical and isotopic systematics of oceanic basalts: implications for mantle compositions and processes. In: Saunders AD, Norry MJ (eds) Magmatism in ocean basins. Geological Society of London, Special Publication 42, pp 313–345

Sun SS, Nesbitt RW (1977) Chemical heterogeneity of the Archaean mantle, composition of the Earth and mantle evolution. Earth Planet Sci Lett 35:429–448

Sun SS, Nesbitt RW (1978) Petrogenesis of Archean Ultrabasic and basic volcanics: evidence from rare earth elements. Contrib Mineral Petrol 65(301):325

Swager CP, Goleby BR, Drummon BJ, Rattenbury MS, Williams PR (1997) Crustal structure of granite-greenstone terrains in the Eastern Goldfields, Yilgarn Craton, as revealed by seismic reflection profiling. Precambrian Res 83:43–56

Tarney J, Windley BF (1977) Chemistry, thermal gradients and evolution of the lower continental crust. J Geol Soc Lond 134:153–172

Tarney J, Dalziel IWD, DeWit MJ (1976) Marginal basin 'Rocas Verdes' complex from south Chile: a model for Archaean greenstone belt formation. In: Windley BF (ed) Early history of the Earth. Wiley, London, pp 131–146

Tatsumoto M (1978) Isotopic composition of lead in oceanic basalt and its implication to mantle evolution. Earth Planet Sci Lett 38:63–87

Taylor SR (1967) The origin and growth of continents. Tectonophysics 4:17–29

Taylor PN, Kramers JD, Moorbath S, Wilson JF, Orpen JL, Martin A (1991) Pb/Pb, Sm-Nd and Rb-Sr geochronology in the Archean craton of Zimbabwe. Chem Geol (Isot Geosci Sect) 87:175–196

Tegtmeyer AR, Kröner A (1987) U-Pb zircon ages bearing on the nature of early Archean greenstone belt evolution, Barberton Moutainland, Southern Africa. Precambrian Res 36:1–20

Terabayashi M, Masuda Y, Ozawa H (2003) Archaean ocean floor metamorphism in the North Pole area, Pilbara Craton, Western Australia. Precambrian Res 127:167–180

Tessalina SG, Bourdon B, Van Kranendonk MJ, Birck JL (2010) Influence of Hadean crust evident in basalts and cherts from the Pilbara Craton. Nat Geosci 3:214–217

Thiemens MH (1999) Atmospheric science – mass-independent isotope effects in planetary atmospheres and the early solar system. Science 283:341–345

Thorne AM, Trendall AF (2001) Geology of the Fortescue Group, Pilbara Craton, Western Australia. Geol Surv West Aust Bull 144:249

Thorpe RI, Hickman AH, Davis DW, Mortensen JK, Trendall AF (1992) U–Pb zircon geochronology of Archaean felsic units in the Marble Bar region, Pilbara Craton, Western Australia. Precambrian Res 56:169–189

Thurston PC (1994) Archaean volcanic patterns. In: Condie KC (ed) Archaean crustal evolution. Elsevier, Amsterdam, pp 45–84

Tice MM, Lowe DR (2006) The origin of carbonaceous matter in pre-3.0 Ga greenstone terrains: a review and new evidence from the 3.42 Ga Buck Reef Chert. Earth Sci Rev 76:259–300

Tingey RJ (1991) The regional geology of Archaean and Proterozoic rocks in Antarctica. In: Tingey RJ (ed) The geology of Antarctica. Oxford University Press, Oxford, pp 1–73

Treloar PJ, Blenkinsop TG (1995) Archaean deformation patterns in Zimbabwe: true indicators of Tibetan style crustal extrusion or not? In: Coward MP, Ries AC (eds) Early Precambrian processes. Geological Society of London special publication 95. Geological Society, London, pp 87–108

Trendall AF (1995) Paradigms for the Pilbara. In: Coward MP, Ries AC (eds) Early Precambrian processes. Geological Society of London special publication 95. Geological Society, London, pp 127–142

Trendall AF, Blockley JG (1970) The iron formations of the Precambrian Hamersley Group, Western Australia. Geological Survey Western Australia Bulletin 119, 365 pp

Trendall AF, de Laeter JR, Nelson DR, Bhaskar Rao YJ (1997) Further zircon U–Pb age of the Daginkatte For-mation, Dharwar Supergroup, Karnataka craton. J Geol Soc India 50:25–30

Trendall AF, Nelson DR, deLaeter JR, Hassler SW (1998) Precise zircon U-Pb ages from the Marra Mamba Iron Formation and Wittenoom Formation, Hamersley Group, Western Australia. Aust J Earth Sci 45:137–142

Trendall AF, Compston W, Nelson DR, De Laeter JR, Bennett VC (2004) SHRIMP zircon ages constraining the depositional chronology of the Hamersley Group, Western Australia. Aust J Earth Sci 51:621–644

Tsomondo JM, Wilson JF, Blenkinsop TG (1992) Reassessment of the structure and stratigraphy of the Early Archean Selukwe nappe, Zimbabwe. In: Glover JE, Ho SE (eds) The Archaean: terrains, processes, and metallogeny: Geology Department (Key Centre), University of Western Australia Geology Department and University Extension publication 22. University of Western Australia, Nedlands, pp 123–135

Turekian K, Wedepohl KH (1961) Distribution of the elements in some units of the Earth's crust. Geol Soc Am Bull 72:175–185

Uysal IT, Golding SD, Glikson AY, Mory AJ, Glikson M (2001) K–Ar evidence from illitic clays of a late Devonian age for the 120 km diameter Woodleigh impact structure Southern Carnarvon basin Western Australia. Earth Planet Sci Lett 192:281–289

Valley JW (2005) A cool early Earth. Sci Am 293(4):58–66

Valley JW (2008) The origin of habitats. Geology 36:911–912

Valley JW, Peck WH, King EM (2002) A cool early Earth. Geology 30:351–354

Valley JW, Cavosie AJ, Fu B, Peck WH, Wilde SA (2006) Comment on Heterogeneous hadean hafnium: evidence of continental crust at 4.4 to 4.5 Ga. Science 312:57–77

Van Kranendonk MJ (2000) Geology of the North Shaw 1:100,000 sheet. Western Australia Geological Survey 1:100,000 Series Explanatory Notes, 89 pp

Van Kranendonk MJ (2010) Three and a half billion years of life on Earth: a transect back into deep time. Geological Survey of Western Australia, Record 2010/21, 1–93

Van Kranendonk MJ (2011) Onset of plate tectonics. Science 333:413–414

Van Kranendonk MJ, Morant P (1998) Revised Archaean stratigraphy of the North Shaw 1:100 000 sheet, Pilbara Craton. Geological Survey Western Australia, Annual Review 1997–1998, pp 55–62

Van Kranendonk MJ, Hickman AH, Smithies RH, Nelson D (2002) Geology and tectonic evolution of the Archaean North Pilbara Terrain, Pilbara Craton, Western Australia. Econ Geol 97:695–732

Van Kranendonk MJ, Webb GE, Kamber BS (2003) Geological and trace element evidence for a marine sedimentary environment of deposition and biogenicity of 3.45 Ga stromatolitic carbonates in the Pilbara Craton, and support for a reducing Archean ocean. Geobiology 1:91–108

Van Kranendonk MJ, Hickman AH, Smithies RH, Williams IR, Bagas L, Farrell TR (2006) Revised lithostratigraphy of Archaean supracrustal and intrusive rocks in the northern Pilbara Craton, Western Australia. Geological Survey of Western Australia, Record 2006/15, 1–57

Van Kranendonk MJ, Smithies RH, Hickman AH, Champion DC (2007a) Paleo-archaean development of a continental nucleus: the east Pilbara terrain of the Pilbara craton. In: Van Kranendonk MJ, Smithies RH, Bennett VC (eds) Earth's oldest rocks. Developments in Precambrian geology 15. Elsevier, Amsterdam, pp 307–337

Van Kranendonk MJ, Smithies RH, Hickman AH, Champion DC (2007b) Secular tectonic evolution of Archaean continental crust: interplay between horizontal and vertical processes in the formation of the Pilbara Craton, Australia. Terra Nova 19:1–38

Van Kranendonka MJ, Ivanica TJ, Wingate MTD, Kirkland CL, Wyche S (2013) Long-lived, autochtonous development of the Archean Murchison Domain, and implications for Yilgarn Craton tectonics. Precambrian Res 229:49–92

van Zuilen MA, Lepland A, Arrhenius G (2002) Reassessing the evidence for earliest traces of life. Nature 418:627–630

Vasconcelos C, McKenzie JA, Bernasconi S, Grujic D, Tien AJ (1995) Microbial mediation as a possible mechanism for natural dolomite at low temperatures. Nature 377:220–222

Vearncombe S, Vearncombe JR, Barley ME (1998) Fault and stratigraphic controls on volcanogenic massive sulphide deposits in the Strelley Belt, Pilbara Craton, Western Australia. Precambrian Res 88:67–82

Versfeld JA, Wilson AH (1992) Nondweni Group. In: Johnson MR (ed) Catalogue of south African stratigraphic units. South African Commission for Stratigraphy, Council for Geoscience, Pretoria, pp 19–20

Viljoen MJ, Viljoen RP (1969a) The geology and geochemistry of the lower ultramafic unit of the Onverwacht Group and a proposed new class of igneous rocks. Geol Soc S Afr Spec Publ 2:55–86

Viljoen RP, Viljoen MJ (1969b) The geological and geochemical significance of the upper formations of the Onverwacht Group. Geol Soc S Afr Spec Publ 2:113–152

Viljoen MJ, Viljoen RP (1969c) Archaean volcanicity and continental evolution in the Barberton region, Transvaal. In: Clifford JN, Gass JP (eds) African magmatism and tectonics. Oliver and Boyd, Edinburgh, pp 29–47

Viljoen MJ, Viljoen RP, Smith HS, Erlank AJ (1983) Geological, textural, and geochemical features of komatiitic flows from the Komati Formation. Geol Soc S Afr Spec Publ 9:1–20

Wacey D (2012) Earliest evidence for life on Earth: an Australian perspective. Aust J Earth Sci 59:153–166

Walker W (1976) Eras, mobile belts and metallogeny. Geol Assoc Can Spec Publ 14:517–558

Walsh M (1992) Microfossils and possible microfossils from the Early Archean Onverwacht Group, Barberton Mountain Land, South Africa. Precambrian Res 54:271–293

Walsh M, Lowe DR (1985) Filamentous microfossils from the 3,500 Myr-old Overwacht Group, Barberton Mountain Land, South Africa. Nature 314:530–532

Wan YS, Liu D, Song B, Wu J, Yang C, Zhang A, Geng Y (2005) Geochemical and Nd isotopic compositions of 3.8 Ga meta-quartz dioritic and trondhjemitic rocks from the Anshan area and their geological significance. J Asian Earth Sci 24:563–575

Watson EB, Harrison TM (2005) Zircon thermometer reveals minimum melting conditions on earliest Earth. Science 308:841–844

Weber W, Scoates RFJ (1978) Archean and Proterozoic metamorphism in the northwestern Superior Province and along the Churchill-Superior boundary, Manitoba. In: Metamorphism in the Canadian Shield. Geological Survey of Canada paper 78-10. pp 5–16

White DJ, Musacchio G, Helmstaedt HH, Harrap RM, Thurston PC, van der Velden A, Hall K (2003) Images of a lower-crustal oceanic slab: direct evidence for tectonic accretion in the Archean western Superior province. Geology 31:997–1000

Whittaker AJ (2001) Components and structure of the Yilgarn Craton, as interpreted from aeromagnetic data. In: Proceedings of fourth international Archaean symposium ext. Abstracts AGSO-geoscience Australia record, vol 37. pp 536–538. This product is released under the Creative Commons Attribution 3.0 Australia Licence. http://creativecommons.org/licenses/by/3.0/au/deed.en

Wilde SA, Valley JW, Peck WH, Graham CM (2001) Evidence from detrital zircons for the existence of continental crust and oceans on the Earth 4.4 Gyr ago. Nature 409:175–178

Wilhelmj HR, Dunlop JSR (1984) A genetic stratigraphic investigation of the Gorge Creek Group in the Pilgangoora syncline. In: Muhling JR, Groves DI, Blake S (eds) Archaean and Proterozoic Basins of the Pilbara, Western Australia: evolution and mineralization potential. University of Western Australia Geology Department and University Extension publication. University of Western Australia, Nedlands, pp 68–88

Wilhelms DE (1987) The geological history of the Moon. US Geological Survey Professional paper 1348

Williams IR (1999) Geology of the Muccan 1:100 000 Sheet. Geological Survey of Western Australia 1:100 000 Geological Series Explanatory Notes

Williams IR (2001) Geology of the Warrawagine 1:100 000 Sheet. Geological Survey of Western Australia 1:100 000 Geological Series Explanatory Notes

Williams IR (2003) Geology of the Yilgalong 1.100 000 sheet. Geological Survey of Western Australia Record

Williams IR, Bagas L (2007) Geology of the Mount Edgar 1:100 000 Sheet. Geological Survey of Western Australia 1:100 000 Geological Series Explanatory Notes

Williams DAC, Furnell RG (1979) A reassessment of part of the Barberton type area. Precambrian Res 9:325–347

Williams IR, Myers JS (1987) Archaean geology of the Mount Narryer region, Western Australia. Western Australia Geological Survey, Report 22, 32 pp

Williams HR, Stott GM, Thurston PC, Sutcliffe RH, Bennett G, Easton RM, Armstrong DK (1992) Tectonic evolution of Ontario: summary and synthesis. In: Thurston PC et al (eds) Geology of Ontario. Ontario Geological Survey special 4. pp 1255–1332

Wilson JF (1973) Granites and gneisses of the area around Mashaba, Rhodesia. Geol Soc S Afr Spec Publ 3:79–84

Wilson JF (1979) A preliminary reappraisal of the Rhodesian basement complex. Geol Soc S Afr Spec Publ 5:1–23

Wilson AH, Carlson RW (1989) A Sm-Nd and Pb isotope study of Archaean greenstone belts in the southern Kaapvaal Craton, South Africa. Earth Planet Sci Lett 96(1–2):89

Wilson JF, Baglow N, Orpen JL, Tsomondo JM (1990) A reassessment of some regional correlations of greenstone-belt rocks in Zimbabwe and their significance in the development of the Archaean craton. In: Third international Archaean symposium, Perth 1990. Geoconferences, Perth, pp 43–44

Wilson JF, Nesbitt RW, Fanning CM (1995) Zircon geochronology of Archaean felsic sequences in the Zimbabwe craton: a revision of greenstone stratigraphy and a model for crustal growth. In: Coward MP, Ries AC (eds) Early Precambrian processes. Geological Society of London special publication 95. Geological Society, London. pp 109–126

Windley BF (1973) Archaean anorthosites: a review of the Fiskenaesset Complex, west Greenland, as a model for interpretation. Geol Soc S Afr Spec Publ 3:319–332

Windley BF, Bridgwater D (1971) The evolution of Archaean low- and high-grade terrains. Geol Soc Aust Spec Publ 3:33–46

Windley BF, Smith JV (1976) Archaean high-grade complexes and modern continental margins. Nature 260:671–675

Wooden JL, Mueller PA (1988) Pb, Sr, and Nd isotopic compositions of a suite of Late Archean igneous rocks, eastern Beartooth Mountains: implications for crust-mantle evolution. Earth Planet Sci Lett 87:59–72

Woodhead JD, Hergt JM, Simonson BM (1998) Isotopic dating of an Archaean bolide impact horizon, Hamersley basin, Western Australia. Geology 26:47–50

Wyche S (2007) Evidence of pre-3100 Ma crust in the Youanmi and southwest terranes and Eastern Goldfields super-terrain of the Yilgarn Craton. In: Van Kranendonk MJ, Smithies RH, Bennett VC (eds) Earth's oldest rocks. Developments in Precambrian geology 15. Elsevier, Amsterdam, pp 113–123

Wyche S, Nelson DR, Riganti A (2004) 4350–3130 Ma detrital zircons in the Southern Cross granite-greenstone terrain, Western Australia: implications for the early evolution of the Yilgarn Craton. Aust J Earth Sci 51:31–45

York D, Layer PW, Lopez Martinez M, Kröner A (1989) Thermal histories from the Barberton greenstone belt, southern Africa. In: International Geological Congress, Washington, DC, p 413

Young GM (1978) Some aspects of the evolution of the Archaean crust. Geosci Man 5:140–149

Young GM, von Brunn V, Gold WEL, Minter DJC (1998) Earth's oldest reported glaciation: physical and chemical evidence from the Archean Mozoan Group (~2.9 Ga). S Afr J Geol 106:523–538

Zahnle K, Sleep NH (1997) Impacts and the early evolution of life. In: Thomas P, Chyba C, McKay C (eds) Comets and the origin of life. Springer, Dordrecht, pp 175–208

Zegers TE, Nelson DR, Wijbrans JR, White SH (2001) SHRIMP U–Pb dating of an Archaean core complex formation and pancratonic strike-slip deformation in the East Pilbara Granite–Greenstone Terrain. Tectonics 20:883–908

Zeh A, Gerdes A, Barton JM (2009) Archean accretion and crustal evolution of the Kalahari Cratonç the zircon age and Hf isotope record of granitic rocks from Barberton/Swaziland to the Francistown Arc. J Petrol 55:933–966

Zeh A, Stern RA, Gerdes A (2014) The oldest zircons of Africa – their U–Pb–Hf–O isotope and traceelement systematics, and implications for Hadean to Archeancrust–mantle evolution. Precambrian Res 241:203–230

Zhao GC, Sun M, Wilde SA, Li SZ (2005) Late Archaean to Palaeoproterozoic evolution of the NCC: key issues revisited. Precambrian Res 136:177–202

Index

A
Abiotic photo-oxidation, 171
Abitibi, 70
Acasta, 14–15
Acasta Gneiss, 10
Accretion, 52
Accretional lapilli, 84, 85
Accretionary prisms, 125
Accretion-differentiation events, 179
Acidity, 159
Adakites, 110, 128, 182
Adamellite, 37, 60, 179
Aerobic eukaryotes, 165
Aerodynamic cavities, 94
Aerodynamic effects, 134
Africa, 23
Agglutinated microkrystite spherules, 147
Agmatites, 30
Airborne magnetic, 98
Airborne magnetic data, 123
Aitken, 11
Akilia, ix, 17, 21, 26, 159
Akilia apatite, 171
Akilia banded ironstones, 171
Alamo breccia, 6
Albedo, 168
Albite porphyries, 61, 62
Aldan Shield, 22
Al-depleted, 55
Algae, 166
Alkali element, 56
Alkaline, 180
Alkaline rocks, 54
Allochtonous thrust block, 11
Alpine, 161
Alpine ophiolite-melange, 124
Altered glass, 95
Al-undepleted, 55
Al-undepleted komatiite, 16
Amino acids, 171
Amitsoq, 64
Amitsoq gneisses, 62, 64, 65
Amphibole, 67
Amphibole fractionation, 66
Amphibolite facies, 33
Amphibolites, 2, 18, 33, 38, 51, 66, 71, 120, 127
Amphibolite screens, 39
Anabar Shield, 22
Anaerobic, 17
Anataxis, 70
Anatectic/metamorphic, 23
Anatectic processes, 180
Anatexis, 15, 67, 69
Ancient Gneiss Complex (AGC), 26, 34, 62, 69, 75, 77
Ancient tonalites, 38, 62, 66
Andalusite-bearing, 127
Andean, 161
Andesites, 71, 128, 179, 180
Andesitic melts, 67
Anhydrous, 67
Ankerite, 85
Anorthositic xenoliths, 33
Anoxic, 17
Anshan Group, 23
Antarctic, xiii
Antarctica, ix, 2
Antarctic Chert Member, 83, 113, 114, 156, 157, 183
Antarctic Creek Member, 112, 115
Apatite, 95

Aplite, 39
Apollo, 4
Apophysae, 38
Arc-continent collisions, 129
Archaean, x, 10, 14
Archaean atmosphere, 21
Archaean atmospheric conditions, 157
Archaean batholiths, 66, 180
Archaean episodes, 182
Archaean geotherms, 54
Archaean granitic crust, 38
Archaean granitoids, 96
Archaean impact, ix
Archaean impact clusters, 167
Archaean mafic-ultramafic, 67
Archaean mantle, 54, 68
Archaean oceans, 161
Archaean precursors materials, 46
Archaean tectonic-thermal history, 180
Archaean tonalite-trondhjemite granodiorite, 69
Archaean zircons, 48
Arc-trench, 178
Arc-trench volcanics, 180
Arenites, 84, 86, 112, 113, 146
Argillite, 117
Argillite-chert sequence, 135
Arnaud River, 16–17
Arsenide, 153
Assimilation, 36, 67
Asteroid bombardment, ix
Asteroid clusters, 177
Asteroid impact, 83
Asteroid impact clusters, 73
Asteroid impact ejecta, 75
Asthenosphere, 54
Atmosphere, 162
Atmosphere-ocean system, 75
Atmospheric CO_2, 161
Atmospheric gases, 1
Atmospheric greenhouse gas, 168
Atmospheric methane, 165
Atmospheric oxygen, 159
Australia, 23
Authigenic, 134
Autochtonous, 87
Autotrophs, 162, 166

B

Bababudan Group, 127
BABI. *See* Basaltic achondrite best initial composition (BABI)
Ball-and-pillow, 146

Baltic/Fennoscandian, xiv
Banded chert, 85
Banded ferruginous chert, 153
Banded gneiss, 123
Banded iron formation (BIF), 3, 10, 15, 20, 21, 117, 156, 162
Banded iron sediments, 171
Banded ironstones, 87, 159
Barberton, ix, 7, 14, 38
Barberton Greenstone Belt (BGB), 5, 7, 30, 38, 56, 75, 77, 88, 95, 102, 113, 114, 125, 156
Barberton Mountain Land, 157
Barberton plutons, 60
Barberton-Swaziland, 64
Barberton tonalites, 64
Barite, 86, 112, 172
Basal conglomerate, 136
Basalt-andesite, 116
Basalt-andesite-dacite-rhyolite, 87
Basaltic achondrite best initial composition (BABI), 46
Basaltic andesite, 67
Basaltic komatiites, 75, 81
Basalts, 81, 151
Basal unconformities, 180
Base Belvue Formation, 92
Basement schists, 37
Basic dykes, 31
Basic granulite, 66
Basic magmas, 67
Basic pods, 67
Bastar craton, 126
Batholiths, 26, 33, 60, 101
Bearooth Mountains, 17
Beartooth, 18
Beartooth-Bighorn Magmatic Zone, 18
BEBI. *See* Bulk Earth Best Initial (BEBI)
Bee Gorge Member, 144, 156, 157
Belingwe belt, 34, 88
Below-wave-base, 94, 145
BGB. *See* Barberton Greenstone Belt (BGB)
Bighorn subprovince, 18
Bilateral motile animals, 169
Biogenic, 171
Biogenic activity, 159
Biogenic productive periods, 162
Biogenic sulfur, 172
Biological activity, 166
Biomolecules, 170
Biotite–tonalite enclaves, 99
Black chimneys, 170
Black Range Dolerite Suite, 103
Black shale, 162, 166

Index

Bombardment, 22
Boninites, 54, 128
Bosmanskop pluton, 66
Boulder deposit, 136
Bouma cycle, 110
Bouma-type cycle, 146
Breccia, 141
Breccia fragment, 143
Brecciated chert, 159
Brockman Iron Formation, 151, 157
Bruno's Band, 156
Bubbles, 147
Budjan Creek Formation, 108
Building blocks of life, 170
Bulawayan, 65
Bulawayan Group, 87
Bulawayo, 37
Bulk Earth Best Initial (BEBI), 46
Bulun, 22

C

Ca-Al inclusions, ix
Calc-alkalic, 69, 70
Calc-alkaline, 22, 62, 70, 87, 110
Calc-alkaline volcanic cycles, 56
Calc-alkaline volcanics, 86
Caledonian, 161
Cambrian explosion of life, 169
Canada, x, 56
Cap carbonate, 168
Capricorn mobile belt, 125
Capricorn orogen, 120
Carawine, xiv
Carawine Dolomite, 138, 139, 140, 141, 143, 156
Carawine Dolomite spherule bearing mega-breccia, 141
Carbon, xiv
Carbonaceous chondrites, ix, 93, 95, 96, 156
Carbonaceous matter, 171, 172
Carbonaceous sediment, 93
Carbonaceous shale, 161
Carbonate, 17, 87, 127, 133, 145, 146, 147
Carbonate-argillite matrix, 144
Carbonated tholeiitic basalts, 56
Carbonates, 105, 131
Carboniferous-Permian boundary, 6
Carbon isotopic compositions, 166
Carbon/Nitrogen ratios, 172
Carbon-oxygen-sulphur cycles, 160–170
Carlindi Granitic Complex, 101
Catastrophic, xiv
C1-chondrite, 149

Cellular structure, 167
Central Pilbara, 159
Chalcophile spinels, 95
Chalcopyrite, 95
Charlevoix, 6
Charnockite, 127
Chemo-lithotropic, 169
Chemosynthetic microbes, 172
Chemotrophic, 2, 167
Chert, 105
Chert-dolomite unit, 141–142
Cherts, 84, 85
Chicxulub, 7
Chicxulub impact structure, 142–143
Chilean Mesozoic granodiorite-greenstone back-arc terrains, 179
Chilimanzi, 37, 65
China, 2
Chingezi tonalite, 87
Chitradurga fault, 128
Chitradurga Group, 127
Chlorite, 81, 85, 94, 139, 140, 147
Chlorite schist, 33
Chondrite, 48
Chondrite-normalized, 149
Chondritic, 15, 95
Chondritic isotopic composition, 47
Chondritic patterns, 95
Chondritic projectile, 155
Chondritic Re/Os, 52
Chondritic unfractionated reservoir (CHUR), 50
Chondritic values, 149
Chotanagpur Gneiss Complex, 126
Chromites, 81, 96
CHUR. See Chondritic unfractionated reservoir (CHUR)
Circum-Pacific, 39, 68
Circum-Pacific arc-trench accretionary systems, 129
Circum-Pacific wedges, 179
Climbing-ripples, 145
Clinopyroxene, 59, 70
Coccoidal bacteria, 171
Collisional orogeny, 130
Colloidal silica, 114
Condensates, 154
Condensation, 139
Condensation spherules, 86
Conglomerates, 12, 78, 84, 86, 115, 125
Contamination, 22
Continental basement, 181
Continental collision and accretion, 120
Continental crust, 46, 120

Continental lithosphere, 54
Continental nuclei,
Contradictory hypotheses, 178
"Cool" early Earth, 13
Coonterunah Group, 113
Coonterunah Subgroup, 182
Corboy Formation, 102, 117
Cordierite-hypersthene-sillimanite-bearing schists, 127
Cordillera, 178
Corruna Downs, 39
Corundum-bearing, 127
Cosmic collision, 1
$^{52}Cr/^{52}Cr$ anomalies, 169
Cr isotopic analyses, 94
Cr-mica, 94
Cross Lake, 40
Cross-rippled quartzite, 127
Cross-ripples, 145
Cr-rich chlorite, 95
Cr-rich rutile, 95
Crust formation, 14
Cryogenian, 168, 169
Crystal fractionation, 58
Crystal mush, 40
Cuddapah Basin, 127
Current-bedded arenite, 95
Current-disturbed turbidite, 147
Cyanobacteria, 165

D
Daciteas, 180
Dacites, 60, 84–86, 128
Dacitic melts, 67
Dakota, 18
Dales Gorge, 152
Dales Gorge Member, 5, 151, 153, 157
Dalmein Pluton, 38
Daly-type batholiths, 36
Decay constant, 45
Deccan plateau, 127
Deep hot biosphere, 3
Deep water shale, 87
Deformed crystallites, 134
De Grey Superbasin, 106
Dehydration, 37
Depleted mantle, 45, 51
Detrital, 2, 11
Detrital zircons, 79, 124, 125
Detritus, 13
Devitrification textures, 95
Devonian, 6
Dharwar, 40

Dharwar Craton, 40, 126, 128, 184
Dharwar greenstone belts, 40
Dharwar Supergroup, 127
Diamictite, 168
Diapiric granitic domes, 98
Diapiric granitoids, 184
Differential partitioning, 48
Diorite, 14, 15, 71
Dioritic, 179
Discrete impact episodes, 73
Dolerite, 112, 113
Dolerite sills, 110, 112
Dolomite, 3, 144, 171
Doornhoek, 78
Doornhoek TTG plutons, 78
Doornhook Pluton, 38
Dresser Formation, 112, 167, 172–174
Drumbeg gneisses, 64
Duffer Formation, 70, 112, 113, 183
Dumbbell-like forms, 153
Dumbbell-shaped spherules, 152
Dwalile, 26

E
Early life, 20, 75
Earth, ix
Earth mantle, 52
East Antarctic, 10, 23
Eastern Ghats, 34
Eastern Ghats mobile belt, 127
Eastern Goldfields, 123
Eastern Goldfields Province, 124
East Fortescue Basin, 175
Eclogite-bearing associations, 130
Eclogites, 66, 67, 71
Ejecta, 86
Ejecta/fallout units, 131
Empirical methodology, 178
Enclaves, 180
Enderbite, 127
English River, 129
Enhanced biological activity, 162
Ensialic, 130
En-sialic granitoids, 130
Ensimatic, 130
Entrained zircons, 181
Episode, 124
Episodic evolution, 51
Episodic uniformitarianism, 179
E-probe, 114
Eratosthenian, 4
Erosion, 86
Errabiddy, 11

Index 229

Eu anomalies, 70
Eu/Eu* anomalies, 70
Eukaryotes, 168
Eutectic, 14
Eutectic K-rich granites, 179
Evaporite, 85, 161
Evaporite deposits, 161
Evolution, 73, 125
Evolutionary theories, 181
Exhalations, 17
Exhalites, 17
Exobiogenic, 170
Extraterrestrial, 152
Extreme hydraulic pressure, 143
Extremophile, 2
Extremophile cyanobacteria, 161
Extremophile microbial communities, 170

F

Fe isotopic composition, 17
Feldspathic arenite, 117
Felsic, ix
Felsic breccia, 81
Felsic granulites, 22
Felsic magmas, 26, 30, 69
Felsic magmatic rocks, 52
Felsic pyroclastic, 117
Felsic tuffs, 85, 86
Felsic volcanics, 78, 81, 87
Felsic volcanic sequences, 106
Fennoscandian/Baltic terrains, 130
Fennoscandian Shield, 130
Fe-oxides, 147
Ferric iron oxidation, 166
Ferroan dolomite, 20
Ferro-gabbro, 18
Ferruginous argillite, 117
Ferruginous chert, 85
Ferruginous sediments, 92
Ferruginous shale, 169
Fig Tree Group, 80, 85, 86, 90, 94, 96, 114, 115
Filamentous, 172
Filamentous microbial remains, 172
Filamentous Prokaryote, 172
Finland, 34, 56
Fiskenaesset region, 67–68
Flat subduction, 182
Flattened peloids, 147
Foreland, 86
Fortescue, xiv
Fortescue Basin, 103, 125, 175
Fortescue Group, 40, 58
Fortescue River valley, 135

Fractional crystallization, 67, 70
Fractionation, 45, 68
Fractionation of the PGE, 134
Fra Mauro, 4
Free atmospheric oxygen, 171
Fuchsite quartzites, 127
Fumaroles, 159
Fusion of hydrogen to helium, 160

G

Gabbro-anorthosite, 18
Gamahuann carbonates, 154
Ganymede, 6
Garnet, 55, 66, 67, 69, 70
Geochemical, 66
Geochemistry, 12
Geodynamic, 13
Geological mapping, 97, 98
Geophysical surveys, 127
Geotectonic theories, 178
Geothermal gradient, 67
Ghoko, 29
Glacial termination, 169
Glaciations, 168–170
Glass-devitrification textures, 110, 112
Glass shards, 153
Global fallout, 139
Gneiss domes, 26
Gneisses, 11, 12, 33
Gneiss-granulite, 68
Gneiss-granulite root zones, 68
Gneiss-granulite terrains, 60, 126
Goethite, 167
Gondwana, 23
Gorge Creek Basin, 110
Gorge Creek Group, 116, 156, 157
Granite-greenstone, 23, 34
Granite-greenstone terrains, 64, 66
Granite laccolith, 116
Granite Mountain, 17
Granites, 23, 39
Granite sheet, 123
Granitic, 15
Granitic crust, 51
Granitic magmas, 15, 26
Granitic tors, 36
Granitoid basement, 33, 34
Granitoid complexes, 130
Granitoid crust, 65
Granitoid magmas, 12
Granitoids, 12, 14, 15, 60, 86
Granodiorites, 23, 30, 36, 39, 60, 71, 72, 87, 179

Granulite, 2, 16, 38, 67
Granulite facies, 16, 22
Granulite-gneiss, 16, 22
Great Falls, 18
Greenhouse gases, 159
Greenland, 17
Greenschist, 127
Greenstone belts, ix, 14, 17, 69, 88
Greenstone-granite terrains, 14
Greenstone-granitoid nucleus, 77, 78
Greywacke, 85
Griqualand West Basin, 131, 157
Griquastad, 154
Guiana, 65
Gwenoro Dam, 27, 30

H
Habitats, 159
Hadean, 14–16, 48
Hadean geotherm, 14
Hamersley Basin, 5, 7, 134, 147, 156, 166
Hardey Formation, 103, 105
Harzburgites, 96, 134, 154
Heany, 37
Heliotropic, 175
Heliotropic stromatolite reefs, 169
Hematite, 167, 175
Hercynian, 161
Hesta siding, 135
Heterogeneous mantle sources, 70
^{176}Hf/^{177}Hf, 43
Hf-isotope, 125
Hf isotopic variability, 48
Hf model ages, 125
HFSE elements, 56
High-amplitude tsunami wave, 143
High-energy current activity, 147
High field strength (HFS) elements, 53
High field strength lithophile (HFSL), 65
High-grade metamorphic, 18, 26, 67
High-K basalts, 105
High-K granites, 65
High-K rocks, 71
High-level intrusions, 66
High-magnesium basalts, 88
High-Mg basalts, 54, 57, 69, 75, 110, 112, 114
High-temperature melting, 180
Himalayan, 161
Holenarsipur, 33
Holenarsipur greenstone belt, 33
Honeyeater Basalt, 102
Hoogenoeg, 78
Hoogenoeg Formation, 83, 84, 88, 92, 113

Hornblende, 66
Horst and graben, 117
Hudson Bay, 16, 129
Hudson Bay terrain, 129
Huronian glaciation, 168
Hydraulic pressures, 142
Hydrodynamic pressures, 94
Hydrogen, 171
Hydrological cycle, 168
Hydrosphere, 161
Hydrospheric conditions, 169
Hydrothermal, 70, 84, 168
Hydrothermal environments, 172
Hydrothermal system, 172
Hydrothermal vents, 171
Hypabyssal, 125
Hypabyssal porphyries, 60
Hyperthermophilic Methano-caldococcus jannaschii, 172

I
Ice-rafted debris, 168
Ilbiana Well, 135
Ilmenite, 154
Imbrium, 4
Impact basins, 96
Impact crater, 139
Impact ejecta, 84
Impact ejecta units, 177
Impact fallout units, 110–114
Impact incidence rates, 96
Impact-triggered tsunami wave, 114
Incompatible elements, 59, 70
India, 2, 23
Indian Peninsular gneiss, 33
Indian Shield, 34
Initial ^{87}Sr/^{86}Sr, 64
Inter-bioherm interstitial debris, 175
Intergranular carbonate, 144
Intermediate-pressure kyanite-sillimanite facies, 127
Intermontane, 86
Internal crystallites, 134
Inter-thrusting, 68
Intra-batholithic, 34
Intra-cratonic rift zones, 125, 182
Intra-formational conglomerate, 94
Intra-oceanic arc setting, 110
Intra-plate basalts, 88
Intra-plate settings, 178
Inward-radiating, 96
Inward-radiating K-feldspar fans, 147, 152
Inyoka Fault, 85

Iridium, 92
Irkut, 22
Iron formation, 87
Iron meteorites, 52
Iron oxide, 138
Iron-oxidizing microbes, 162, 166
Iron-rich dolomite, 85
Ironstones, 5
Island-arc andesites, 71
Island-arcs, 68
Island continents, 35
Isochron age, 48
Isotope anomalies, 5
Isotopic age zonation, 119
Isotopic mantle signatures, 35
Isotopic parent-daughter, 43
Isotopic parent-daughter decay constants, 43
Isotopic tracers, 159
Isua, ix, 5, 22, 51
Isua greenstone belt, 159
Isua supracrustal, 5, 15, 18
Isua supracrustal belt, 18
Itsaq, xiii
Itsaq Gneisses, 18–22

J
Jack Hills, ix, 12, 13, 125
Jaspilite, 112, 113
Jeerinah, xiv
Jeerinah Formation, 103, 105, 141, 156, 157
Jeerinah Impact Layer, 157

K
Kaapvaal, 5, 39
Kaapvaal Craton, 56, 61, 73, 88, 90, 92–94, 113, 181
Kaap Valley, 72
Kaap Valley Pluton, 38
Kaap Valley suite, 79
Kalahari Craton, 96
Kalgoorlie-Norseman greenstone, 56
Kaluga, 6
Kangaroo Caves Formation, 101, 102, 108, 116
Karelia granite-greenstone province, 130
Karnataka, 26
Karratha, 106
Karratha block, 102
Kelly Belt, 108
Kemp Land, 23
K-feldspar, 65, 76, 86, 133–136, 144, 147, 153, 154

K-granites, 18
Kimberley, 75
Kinetic control, 170
Kitoy, 22
Komati Formation, 81, 83, 88, 90
Komatiite basalts, 83, 84
Komatiites, 38, 52, 55–57, 81, 84, 88, 92, 94, 95, 151
Komatiitic-basaltic volcanism, 125
K-rhyolites, 60
K-rich adamellite, 62
K-rich granites, 36, 69, 180
K-rich granitic magmas, 34
K-rich granitoids, 26, 33, 66, 88
K-rich intrusions, 66
K-rich magmas, 60
K-rich rhyolite, 70
Kromberg Formation, 84, 90
Kromberg suite, 78
Kromberg Syncline, 84
K-T boundary, 96
Kunagunarinna Formation, 101, 116
Kurrana, 106
Kurrana block, 102
Kurrana Terrain, 109
Kurrawang conglomerate, 64
Kuruman, 7
Kuruman Banded Iron Formation, 131
Kuruman Iron Formation, 154
Kuruman spherule layer, 134
Kylena Formation, 105

L
Labrador, 5, 26, 40
La Grande, 129
Lake Baikal, 22
Lalla Rookh, 14
Lamina fabrics, 175
Laminated siltstone, 92
Lapilli, 84, 85
Laramide, 17
Large asteroid impacts, 102
Large Igneous Provinces, 182
Large ion lithophile, 39
Large ion lithophile elements (LIL), 35
Laser ablation, 96
Laser ICPMS, 114
Late Archaean, 22
Late Eocene, 6
Late heavy bombardment (LHB), ix, 73
Late K-granite, 87
Late-Neoproterozoic, 169
Lateral accretion, 184

Laws of complexity, 170
Leilira Formation, 107, 117
Leucocratic trondhjemites, 71
Lewisian gneisses, 64
Lewisian granulites, 67
Light rare earth elements (LREE), 16
LIL-element, 179
Limestone, 88
Liquid immiscibility, 84
Lithic, 14
Lithic fragments, 93
Lithophile elements, 67
Lithophile spinel, 95
Lithospheric mantle, 59
Lit-par-lit injection, 30
Lochiel Granite, 38
Lochiel Plateau, 76
Lord Howe Island, 182
Low-energy conditions, 94
Lower Bulawayan Group, 87
Low-K tholeiite-andesite-dacite-rhyolite assemblages, 68
Low-K tholeiite-Na dacite, 69
Low-pressure andalusite-sillimanite, 127
Low-pressure melting, 54
Low Titanium basalts, 4
Low velocity zone, 54
LREE-depleted mantle, 50
LREE enrichment, 51
LREE/HREE fractionation, 53
Lu-Hf system, 17, 46–48
Lunar Mare, ix, 1

M
Macroband, 152
Maddina Formation, 103, 105
Mafic, ix, 10, 18, 36
Mafic crust, 67
Mafic dykes, 85
Mafic dyking, 16
Mafic/felsic, 114
Mafic lapilli tuff, 85
Mafic schlieren, 31
Mafic-ultramafic, 22, 25–34, 36, 68, 90
Mafic-ultramafic crust, ix, 68, 184
Mafic-ultramafic magmatic activity, 125
Mafic-ultramafic sequences, 68, 182
Mafic-ultramafic volcanic sequences, 106
Mafic-ultramafic volcanism, 177
Mafic xenoliths, 62
Magmas, 22, 50, 67
Magmatic arc, 128
Magmatic cycles, 77

Magmatic diapirism, 26
Magmatism, 119
Magnetic and gravity lineaments, 178
Magnetic field, 1
Magnetic/structural zones, 123
Maitland River Supersuite, 110
Malene Supracrustals, 40
Mallina Basin, 99, 110
Manicouagan, 7
Maniitsoq, 6, 7, 179
Mantle, 14, 46, 64
Mantle-crust interaction, 45
Mantle diapirs, 54
Mantle evolution models, 45
Mantle/melt partition, 43
Mantle plumes, 54, 60, 178, 182
Mantle rebound, 184
Mantle wedges, 179
Mapepe Formation, 85, 90, 92, 114
Marble, 34
Marble Bar greenstone belt, 101
Mare, 3
Mare basins, 4
Mare-like crust, 180
Marine carbonates, 46
Marine environments, 86
Marker chert, 117
Marra Mamba Iron Formation, 135, 156
Mashaba, 26, 37
Mass balance calculations, 154, 169
Mass-independent fractionation of sulphur isotopes, 169
Mass independently fractionated sulfur, 172
Matopos, 37
Maynard Hills, 12, 125
Megabreccia, 102, 108, 115, 116, 141–143
Megacrystic anorthosites, 120
Mega-tsunami effects, 142
Mega-tsunamis, 131
Mega-xenoliths, 26
Meghalaya craton, 126
Melange, 39
Melange-type, 179
Melt/crystal equilibria, 46
Melt fragments, 139
Melting events, 184
Mendon Formation, 85, 94, 102, 114, 115
Mesozoic-Cenozoic circum-Pacific belts, 179
Metabolic network, 170
Metabolic pathways, 170
Metamorphic grade, 75
Metamorphism, 15
Metapelite, 127
Metasomatism, 171

Index 233

Meta-volcanic, 20
Meteorites, ix, 10, 170
Meteoritic, 48, 95, 134, 153
Meteoritic origin, 94
Methane, 168
Methanogenic microbial activity, 168
Methodological impasse, 178
Mica schist, 87
Microbes, 172
Microbial, 17
Microbial activity, 85, 171
Microbial habitat, 168
Microbialites, 175
Microbial life, 22
Microbial mats, 174
Microbial methanogenesis, 161, 162, 166
Microbial photoautotrophs, 171
Microbial reactions, 166
Microbial remains, 172
Microbial sulfate reduction, 172
Microbial sulphur metabolism, 162, 166
Micro-continental cratons, 129
Micro-continental terrains, 129
Micro-continents, 22
Micro-cratons, 37, 68
Microcrystalline quartz, 92
Microfossils, 167, 172
Microkrystites, 86, 92, 94, 139, 142, 143, 144, 169
Microkrystite spherules, 95, 133, 135, 139, 142, 143, 147, 154, 183
Microlites, 138, 147
Microscopic microbial lamina, 174
Micro-spherules, 153
Microtektite glass fragments, 143
Microtektites, 136, 140, 142, 144
Middle Marker, 83
Mid-ocean ridge basalts (MORB), 45, 51, 56
Migmatites, 36, 37, 62, 87
Migmatization, 37
Miller–Urey, 170
Minnesota, 64
Minnesota-Ontario, 64
Minnesota River gneisses, 129
Minnesota River Valley, 129
Minnesota River Valley terrain, 129
Model Sm/Nd ages, 50
Modern circum-Pacific domains, 179
Modern mat-dwelling cyanobacteria, 171
Modern oceanic crust, 182
Molasse, 110
Molasse-like conglomerates, 124
Montana, 17
Mont d'Or gneiss, 65

Monteville, 7
Monteville Formation, 131, 157
Monteville Spherule Layer, 133
Monzogranite, 99, 110
Moodies Group, 80, 86
Moon, 10
MORB. *See* Mid-ocean ridge basalts (MORB)
Mosquito Creek Basins, 99, 103, 110
Mount Ada Basalt, 56, 83, 112, 156, 157, 183
Mountain Land, 38
Mount Edgar, 39, 65
Mount Edgar batholith, 32, 33, 39
Mount McRae Shale, 156
Mount Narryer, 11–14
Mount Roe Basalt, 103
Mozaan Group, 168
Mt Narryer, 10, 125
Mt Narryer-Jack Hills, 10
Muccan Batholith, 33
Multicellular animals, 169
Multi-disciplinary studies, 178
Multiple impacts, 114, 184
Multiple spherule ejecta units, 97
Multistage accretion, 55
Murchison greenstone belt, 34
Mushandike Granite, 37

N
Nain Complex, 26
Na keratophyres, 124
Napier Complex, 23
Na-rich, 36
Na-rich dacite, 179
Na-rich granitoids, 62
Na-rich rhyolite, 179
Narryer, 11, 12
Narryer terrain, 124
Natale, 30
Natural laws, 178
Natural phenomena, 178
Nd isotopic data, 18
^{143}Nd/^{144}Nd, 43
Near-Earth asteroids (NEA), 6
Near-liquidus tonalitic magma, 30
Nectarian, 1
Negri and Louden volcanics, 110
Nelshoogte, 38, 72
Neptunian sandstone dykes, 117
Ni-arsenide, 153
Nickol River Formation, 109
Nilgiris, 40
Ni-metal, 153
Nimingarra Iron Formation, 156

Ni-oxide, 153
Ni-rich chromites, 95
Ni-rich spinels, 95
Ni-sulphide, 153
Non-biogenic, 171
Norseman-Wiluna Belts, 123
North Atlantic Craton, 65
North China, xiii
Northern China craton, 22
Northern Superior Province, 129
North Pole dome, 173
North Star Basalt, 56
Nuggihalli, 33
Nuk gneisses, 64
Nuvvuagittuq greenstone belt, 17

O

Oakover Valley, 141
Oceanic anoxia, 159
Oceanic crust, 96, 179, 180
Oceanic-like, 25
Oceanic-like crustal zones, 119
Oceanic plagioclase-rich, 62
Ocean island basalts, 51
Octahedral chromites, 95
Octahedral Ni-chromites, 96
Older granitic crust, 67
Olekma, 22
Olistoliths, 117
Olistostrome, 102, 108, 116, 117
Olivine, 81, 95, 144
Onot, 22
Ontario TTG, 64
Onverwacht Anticline, 38, 84
Onverwacht Group, 30, 34, 41, 61, 62, 80, 88, 94, 96, 114, 115, 157, 171, 181
Ophiolite-granodiorite complexes, 68
Ophiolite-melange wedges, 119
Ophiolites, 39, 54, 130, 179
Ophiolites-turbidite-flysch association, 180
Oregon, 18
Organic molecules, 161
O_2-rich atmosphere, 162, 166
Orogenic belts, 69, 161
Orogenic clastic sedimentation, 94
Orthogneisses, 26, 62
Orthopyroxene, 54, 59
Orthopyroxene-bearing magmatic rocks, 127
Osceli, 84
$^{187}Os/^{188}Os$, 43
Otish Basin, 6
Ovoid spherules, 92
Oxford–Stull, 129

Oxidation, 165
Oxidation of ferrous to ferric iron, 169
Oxide, 153
Oxygen, xiv
Oxygen capture, 166
Oxygen isotopes, 13
Oxygen-poor Archaean atmosphere, 169
Oxygen-poor atmospheric and hydrospheric conditions, 167
Ozone, 21, 162
Ozone depletion, 167

P

Paddy Market Formation, 102, 156
Palaeo-biological, 131
Paleomagnetic, 41
Paleo-Proterozoic, 22
Palimpsest, 138
Pan-African, 23
Panorama Formation, 112
Paraburdoo, xiv
Paraburdoo impact spherule unit, 144
Paraburdoo Member, 144
Paraburdoo Spherule Layer, 134, 144
Paraconformities, 7
Parasitic entities, 170
Partial melts, 43
Pd/Ir ratios, 134
Pegmatite, 39
Pegmatitic, 38
Pelitic schist, 34
Peloids, 146
Peninsular gneiss, 36
Peptides, 171
Peridotite, 58
Peridotitic basalt, 83
Peridotitic komatiites, 54, 57, 69, 75–76, 81, 85, 180
Peritidal carbonate platform, 175
Petrogenetic processes, 179
PGE abundances, 55, 154
Phanerozoic batholiths, 180
Phanerozoic ophiolite-granodiorite complexes, 68
Phlogopite, 144
Phlogopite crystallites, 144
Photoautotrophs, 172
Photo-ferrotropic oxidation, 169
Photolysis, 161
Photosynthesis, 2, 168
Photosynthesizing colonial prokaryotes, 162
Photosynthetic oxygen release, 165
Phototrophic bacterial processes, 167

Index

Phytoplankton, 168
Pieterburg, 75
Pikwitonei belt, 40
Pilbara, ix, 5, 26
Pilbara Craton, x, 7, 26, 33, 56, 101, 110, 133, 134, 175, 181
Pilbara greenstone belts, 53
Pilbara Supergroup, 101, 108
Pilbara trondhjemites, 64
Pillow basalts, 20, 84
Pillow lavas, 22, 134, 171
Pincunah Hills Formation, 116, 117
Plagioclase, x, 45, 59, 66, 70, 95, 144
Planar deformation features, 5, 139, 140
Plate tectonics, 68, 161, 178, 182
Plate tectonic theory, 178
Platinum group elements, 52
Plume tectonic processes, 125
Plutonic, 2
Plutonic granitic magmatism, 101
Plutonic units, 101
Plutons, 78
Polymerizing agent, 171
Polymerizing glycine, 171
Pongola Supergroup, 168
Pontiac, 129
Popigai, 7
Porphyritic, 38
Porphyritic gneiss, 66
Porphyritic textures, 98
Post-kinematic granites, 40
Post-tectonic, 36, 86
Post-tectonic granites, 70, 125
Pre-conceived uniformitarian assumption, 178
Prehnite-pumpellyite, 125
Primitive mafic-ultramafic SIMA crust, 125
Primitive mantle, 45, 51
Primordial cells, 170
Primordial solar system, 5
Primordial soup, 170
Projectile, 156
Prokaryotes, 168
Prokaryotic autotrophs, 170
Proterozoic, 17, 46
Proto-cells, 170
Protoliths, 14
Provenance, 14
Proxies, 159
Proxy studies, 161
Pseudomorphs, 95, 144
Pyramid Hills Formation, 102
Pyrite, 95, 134, 172, 175
Pyritic shale, 105
Pyroclastics, 70, 87, 117, 153
Pyrolite, 154
Pyroxene crystallites, 144
Pyroxenes, 45, 46, 81, 95

Q

Quartz, 36, 76, 147
Quartz-dioritic, 18
Quartz-filled vesicles, 138
Quartzite clasts, 12
Quartzites, 11, 17, 34, 87, 124, 125
Quartz monzonites, 66
Quench chromites, 95
Quench-crystallization, 110, 112
Quench-textured, 81, 96
Quench-textured pyroxene, 84
Quench textures, 94, 96, 113
Quetico, 129

R

Racemic abiogenic synthesis, 170
Radiating fans, 134
Radiogenic, 15
Radiometric surveys, 98
Rajasthan (Bundelkhand) craton, 126
Raman imagery, 168
Rayner Province, 23
Rb-Sr, 43
Recumbent folding, 33
Red Sea rift, 182
REE concentrations, 51
REE fractionation, 70
REE patterns, 66
Regal Formation, 109
Regal Terrain, 106
Regolith, 123
Reivilo, 7
Reivilo Formation, 131, 157
Reivilo spherule layer, 133–134
Reliance Formation, 88
Re-Os, 43
Re-Os model ages, 60
Re-Os system, 51–52
Residual Lu-rich mantle, 48
Resorbed phenocrysts, 98
Rhodesdale batholith, 31
Rhyolites, 110, 116
Rift-related basins, 106
Rift valleys, 103
Rift volcanism, 124
Rip-up clasts, 134
Rodinia, 168
Root zones, 37

Roy Hill, 137, 141, 156
Rutile, 92, 94

S

Salisbury kop granite, 38
Sandspruit Formation, 79, 81
Sandur, 33
Sanukitoid, 71
Sargur Group, 26, 128
Scanning Electron Microscopy, 114
Schoengezicht Formation, 90
Scientific progress, 178
Sea of granite, 36
Seawater-like REE and Y signatures, 159
Seawater sulfate, 172
Sebakwian greenstones, 87
Sebakwian Group, 31, 87, 181
Sedimentary environments, 98
Selukwe, 26, 29, 37
Semi-continental environments, 97
Semi-continental nuclei, ix
Sericite, 90, 134, 147
Serpentine, 84
Sesombi, 37, 65
Shale, 131
Shallow water environment, 134
Shamvaian Group, 88
Shaw, 65
Shaw batholiths, 31, 39
Shaw Granitic Complex, 99
Sheba Formation, 85, 90
Shocked quartz, 96, 139–140
Sholl Terrain, 106, 110
Shoshonite, 179
Siberian, xiii
Siberian Craton, 22
Siderophile elements, 52, 95, 154
Siderophile trace element patterns, 134
Sierra Nevada, 179
Siliceous, 64
Siliceous granulite, 66
Silicification, 172
Silicified argillite, 112, 113
Silicified sediments, 172
Silja, 6
Siltstone, 88, 105, 146
SIMA crust, 109
Singhbhum craton, 126
Sinoidal structures, 175
Slave Province, ix, 14, 70
Sm-Nd system, 48–51
Snowball Earth, 168
Soanesville Basin, 106

Soanesville Group, 102, 108, 117
Solar system, ix, 10
Solidus, 67
Somabula, 30, 65
Songimvelo suite, 78
South Africa, 56, 66, 113
Southern Africa, 2
Southern Cross, 34
Southern Cross Domain, 125
Southern India, 26, 36, 40
South Indian, xiv
South Indian Shield, 126
South Pole, 11
Southwest Greenland, 5, 21, 68
Southwest Yilgarn Craton, 124
Sphericity, 113
Spherule-bearing units, 147
Spherule Marker Bed, 134, 143, 157
Spherules, 5, 90, 94, 114
Spinel, 95
Spinel dunite, 67
Spinel pyroxenite, 67
Spinifex-textured, 81
Split Rock Supersuite, 106, 110
$^{87}Sr/^{86}Sr$, 43
Steynsdorp, 72
Steynsdorp anticline, 81
Steynsdorp pluton, 78
Steynsdorp suite, 78
Stilpnomelane, 134, 153, 154
Stilpnomelane-rich spherules, 153
Stolzburg, 30, 38
Stolzburg plutons, 38, 78
Stratosphere, 162
Strelley Granite, 117
Strelley Group, 182
Strelley Pool Formation, 107
Stromatolite-like, 173
Stromatolites, 162, 168
Structural lineaments, 178
Subduction, x, 68, 161
Subduction-accretion tectonic models, 125
Subduction-related, 130
Subduction zones, 182
Sub-horizontal shears and thrusts, 182
Submarine alkaline hydrothermal vents, 170
Submarine fumaroles, 161
Submarine hydrothermal vents, 170
Sub-solidus temperatures, 67
Sudbury, 5, 6, 179, 180
Suess-type batholiths, 36
Sulfide minerals, 172
Sulphate, 167
Sulphide, 116, 153

Index

Sulphur, xiv
Sulphur isotopes, 169
Sulphur isotopic analyses, 161
Sulphur Springs Group, 101, 102, 108, 115, 117, 157
Superior, xiv
Superior Province, 16–17, 39, 56, 67, 68, 70, 129, 166
Supracrustals, 2, 5, 26, 33, 36, 37, 128
Supracrustal xenoliths, 33
Svecofennian orogeny, 130
Sveco-Norwegian-Grenvillian and Caledonian orogenies, 130
Swaziland, 26, 34, 69, 75
Swift Tuttle, 6
Syenogranite, 110
Synkinematic, 129
Syntectonic magmas, 60
Synthesis of acetate, 171
Synthesis of biomolecules, 170
Synvolcanic plutons, 87

T

Tachylite, 153
Talga-Talga Subgroup, 32, 39
Tambourah greenstone belt, 33
Tasman Sea, 182
Tear-drop shaped, 134
Tectonic environments, 68
Tectonic models, 179
Tectono-stratigraphic suites, 78
Tectono-thermal events, 129
Ternovka, 6
Terrestrial, 12
Terrigenous debris, 95
Teton, 17
Theespruit, 30, 38, 72, 78
Theespruit Formation, 41, 182
Theespruit Pluton, 38
Thermal stress, 172
Tholeiitic basalts, 54, 56, 59, 81, 84, 114
Thrusting, 33
Tobacco Root Mountains, 17
Tokwe basement gneisses, 87
Tokwe gneiss terrain, 87
Tokwe River gneiss, 37
Tokwe terrain, 86
Tonalite, 36, 60, 71, 72, 87
Tonalite-greenstone crust, 69
Tonalites, 23, 60, 65
Tonalite-trondhjemite, 26, 34
Tonalite-trondhjemite batholiths, 68
Tonalite/trondhjemite crust, 70

Tonalite-Trondhjemite-Granodiorite (TTG), ix, 14, 22, 23, 34, 60, 65, 71, 127, 180, 181
Tonalitic, 66, 179
Tonalitic diapirs, 38
Tonalitic gneiss, 37
Tonalitic magmas, 66
Tonalitic orthogneiss, 17
Toodyay-Lake Grace Domain, 124
Trace element geochemistry, 60
Transcurrent and thrust deformation, 180
Trans-Hudson, 18
Transvaal, 5
Transvaal Basin, 7, 144, 154
Transvaal Group, 157
Tremolite, 81, 85
Trichomic cyanobacterium-like microorganisms, 172
Trondhjemites, 17, 35, 36, 39, 60, 64, 65, 71
Trondhjemitic, 66, 179
Tsunami, 94, 140
Tsunami effects, 136
Tsunami wave, 143
TTG, 25, 26, 29, 34, 53
TTG batholiths, 67
TTG magmas, 39
TTG suite, 67
Tumbiana Formation, 103, 105
Tungsten, 5
Turbidite arenite, 85
Turbidites, 84, 144
Two-stage mantle melting processes, 179

U

Uivak, 64
Uliak, 26
Ultramafic, ix, 114
Ultramafic ash, 85
Ultramafic flows, 85
Ultramafic-mafic volcanics, 61
Ultramafic residues, 67
Ultramafic xenoliths, 36, 55
Ulundi Formation, 85, 156
Umuduha suite, 79
Unconformably, 87
Unconformities, 7, 106, 107, 114, 180, 182
Uniformitarian, x, 68, 124
Uniformitarian doctrine, 178
Uniformitarianism, 178
Uniformitarian models, 177
$^{238}U/^{204}Pb$, 43
U-Pb-Th isotopic systems, 43
Upernavik Supracrustals, 40

Upper greenschist facies, 125
Upper mantle, 45
U-Th-Pb isotopes, 43–45
UV-photolysis, 172
UV radiation, 2, 162, 165
UV-triggered photo-chemical reactions, 167
UV-triggered reactions, 162

V
μ value, 45
Vanadium, 55
Varioles, 84
Variolite, 84
Variolitic spherulitic textures, 98
Vendian, 167
Vertical tectonics, 68, 184
Vesicles, 84, 96, 147
Volatile PGE, 139
Volatiles, 39, 84
Volcanic breccia, 20, 84
Volcaniclastic, 84, 86
Volcaniclastic breccia, 84
Volcaniclastic rocks, 112
Volcanic sulfur, 172
Volcanic tuff, 127
Volcanic varioles, 117
Volcanism, 1
Volcanogenic, 116
Vredefort, 6, 179

W
Warburton, 7
Warburton twin impacts, 6
Warrawoona, 25
Warrawoona Group, 34
Water, 67
Wawa–Abitibi terrains, 129
Wawa granite-greenstone terrain, 129
Weathering-sequestration of CO_2, 161
Western Australia, ix, 2, 66
Western Dharwar Craton, 128
Western Gneiss Terrain, 124
Western Pilbara, 39

Western Wabigoon, 129
West Hudson Bay, 17
West Pilbara, 56
West Transvaal Basin, 125
Whim Creek Group, 70, 110
Whitewater, 37
Whundo Group, 103, 106, 109, 110
Wind river, 17, 18
Winnipeg River and Marmion terrains, 129
Wittenoom Formation, 140, 144, 156, 157
Wittenoom Gorge, 152
Woodie Woodie, 141
Woodleigh, 6, 7
Wyman Formation, 70
Wyoming, 17, 18
Wyoming Craton, 17, 18

X
Xenocrystic, 1, 2, 25
Xenocrystic zircon, 17
Xenocrysts, 5, 60, 87, 145
Xenolith chains, 30
Xenoliths, 26, 30, 31, 33, 39, 99, 180
Xenolith swarms, 26

Y
Yampire Gorge, 152
Yeelirrie domain, 125
Yeelirrie terrain, 124
Yilgarn, 34, 56, 64
Yilgarn Block, 34
Yilgarn Craton, x, 39, 56, 67
Yilgarn granites, 65
Youanmi, 11

Z
Zeederbergs Formation, 88
Zimbabwe, 26, 27, 37, 39, 56
Zimbabwe Craton, 26, 37, 40, 65, 86, 88
Zircons, 2, 10, 12, 14, 23, 85, 86, 99, 113, 124
Zircon xenocrysts, 124